RAILWAY ENGINEERING

For my parents, Sophia and Aristide

Railway Engineering

V.A. PROFILLIDIS
Associate Professor,
Section of Transportation
Democritus Thrace University
Greece

Avebury Technical
Aldershot • Brookfield USA • Hong Kong • Singapore • Sydney

© V. Profillidis 1995

All rights reserved. No part of this publication may be reproduced, stored in a retrieval system, or transmitted in any form or by any means, electronic, mechanical, photocopying, recording or otherwise without the prior permission of the publisher.

Published by
Avebury Technical
Ashgate Publishing Limited
Gower House
Croft Road
Aldershot
Hants GU11 3HR
England

Ashgate Publishing Company
Old Post Road
Brookfield
Vermont 05036
USA

British Library Cataloguing in Publication Data

Profillidis, Vassilios
   Railway Engineering
   I. Title
   625

ISBN 0 291 39828 6

Library of Congress Catalog Card Number: 95-76120

Printed in Great Britain at the University Press, Cambridge

Contents

Preface by Ph. Roumeguère	XVII
Foreword	XIX

1. Railways and Transport ... 1

- 1.1. Evolution of the railways ... 1
  - 1.1.1. Historical outline ... 1
  - 1.1.2. The golden age of the railways ... 2
  - 1.1.3. Railways and other competing transportation means ... 2
  - 1.1.4. The evolution of railways' organization ... 4
- 1.2. Rail transport characteristics ... 4
- 1.3. Evolution of railway traffic ... 5
  - 1.3.1. Participation of private and public transport modes in the organization of the transportation system ... 5
  - 1.3.2. Share of rail in the transport market ... 8
  - 1.3.3. Evolution of rail passenger traffic ... 8
  - 1.3.4. Evolution of rail freight traffic ... 9
  - 1.3.5. Railway networks and transportation activity in European countries ... 10
  - 1.3.6. Forecasts for the evolution of the transport market ... 12
- 1.4. The dual nature of the railways: business and technology ... 13
  - 1.4.1. Weaknesses inherited by the railways from the past ... 13
  - 1.4.2. Comparative advantages of the railways ... 14
  - 1.4.3. Development strategy and measures to restructure the railways ... 15
- 1.5. High-speed trains ... 16
  - 1.5.1. The application of high-speeds in railways ... 16
  - 1.5.2. Impact of high speeds on the reduction of rail travel times ... 17
  - 1.5.3. High speeds and the increase of traffic ... 18
  - 1.5.4. The European high-speed rail network ... 18
  - 1.5.5. Technical features of high-speed railway lines ... 18

1.6.	The Channel Tunnel Project	21
	1.6.1. Project description	21
	1.6.2. Travel times and forecasted demand	22
1.7.	Aerotrain and Maglev	23
	1.7.1. The Aerotrain	23
	1.7.2. Magnetic levitation trains (Maglevs)	23
1.8.	Other transportation services with good prospects for the railways	24
	1.8.1. Urban rail services	24
	1.8.2. Combined transport	25
	1.8.3. Bulk loads	27
	1.8.4. Integrated rail transport and logistics	27
1.9.	International railway institutions	28
	1.9.1. International Railways Union (UIC)	28
	1.9.2. European Conference of Ministers of Transport	30
	1.9.3. Community of Railway Networks	30
	1.9.4. United Nations Economic Committee for Europe	30
	1.9.5. Research Institute	30

2. The Track System — 32

2.1.	Areas of railway engineering: track, traction, operation	32
2.2.	The components of the track system	33
2.3.	Track on ballast and concrete slab	35
2.4.	Track gauge	35
2.5.	Load per axle and traffic load	36
	2.5.1. Load per axle	36
	2.5.2. Traffic load	37
2.6.	Sleeper spacing	38
2.7.	The wheel-rail contact	38
2.8.	Transverse wheel oscillations along the rail	40
2.9.	Rail mounting angle on sleeper	43
2.10.	Load gauge	43
2.11.	Forces generated by rail vehicle movement - static and dynamic analysis	46
	2.11.1. Forces generated	46
	2.11.2. Static and dynamic analysis	47
2.12.	Influence of the forces generated on passenger comfort	48
2.13.	Construction cost of a new railway line	48

3. Railway Subgrade — 50

3.1.	The importance of the railway subgrade on track quality and its functions	50

	3.2.	Geotechnical analyses and soil classifications	51
		3.2.1. Analytical geotechnical study	51
		3.2.2. Geotechnical classifications of soils	51
	3.3.	Hydrogeological conditions	53
	3.4.	Classification of the railway subgrade	54
	3.5.	Mechanical characteristics of the subgrade	55
	3.6.	The formation layer	56
	3.7.	Impact of traffic load on the subgrade	57
	3.8.	Impact of maintenance conditions on the subgrade	57
	3.9.	Fatigue behaviour of the subgrade	59
	3.10.	Frost protection of railway lines	60
		3.10.1. Frost index	60
		3.10.2. Frost foundation thickness	61
		3.10.3. Frost protection methods on existing tracks	61
	3.11.	Track subgrade in trenches and on embankments - slope gradients	62
		3.11.1. Subgrade in trench	62
		3.11.2. Subgrade on embankment	63
	3.12.	The reinforced soil technique in railway engineering	64
	3.13.	Geotextiles in railway subgrades	65

4. Mechanical Behaviour of Track   68

	4.1.	Analysis of vertical phenomena - track coefficients	68
		4.1.1. Definitions - symbols	68
		4.1.2. Track coefficients	68
	4.2.	Approached elastic analysis of vertical effects - Zimmermann's method	70
	4.3.	Accurate static analysis of vertical effects - Finite-element method	74
		4.3.1. Advantages and procedure of the finite-element method	74
		4.3.2. Limit conditions	75
		4.3.3. Stress-strain relationship	75
		4.3.3.1. Case of ballast and subgrade	76
		4.3.3.2. Case of rail and sleeper	77
		4.3.4. Numerical calculation procedure	77
		4.3.5. Determination of the mechanical characteristics of the various materials	78
		4.3.6. Stress and strain in the track-subgrade system according to the finite-element method	79
		4.3.7. Distribution of wheel load along successive sleepers	81
		4.3.8. Sleeper elastic line	81
	4.4.	Dynamic analysis of the track-subgrade system	82

Railway Engineering

4.5.	Additional dynamic loading	84
4.6.	Design of the track-subgrade system	86
4.7.	Vibrations from rail traffic	86
	4.7.1. Origin of rail vibrations	86
	4.7.2. Relation of rail noise level to speed	87
	4.7.3. Damping of rail noise in relation to distance	87
	4.7.4. Noise level in relation to infrastructure type	87
	4.7.5. Noise levels in high speeds	88
	4.7.6. Noise level standards	88

5. The Rail — 89

5.1.	Rail profiles	89
5.2.	Choice of rail section	90
5.3.	Rail steel grade, mechanical strength and chemical composition	92
	5.3.1. Mechanical strength	92
	5.3.2. Chemical composition	92
	5.3.2.1. Carbon	92
	5.3.2.2. Manganese	92
	5.3.2.3. Chromium	92
	5.3.2.4. Chromium-Manganese	92
	5.3.2.5. Equivalent carbon percentage	93
	5.3.3. Hard steel grades	93
5.4.	Stress analysis of rail	93
	5.4.1. Stresses at wheel-rail contact	94
	5.4.2. Bending stresses of the rail on the ballast	95
	5.4.3. Bending stresses of the rail head on the rail web	95
	5.4.4. Stresses caused by thermal effects	95
	5.4.5. Plastic stresses	96
5.5.	Analysis of the mechanical behaviour of the rail by the finite-element and the photoelasticity methods	97
5.6.	Rail fatigue	98
	5.6.1. Fatigue curve and Miner's rule	98
	5.6.2. Dang Van's rail fatigue criterion	99
	5.6.3. Evolution of an internal discontinuity	100
5.7.	Rail defects	101
	5.7.1. Tache ovale	102
	5.7.2. Horizontal cracking	102
	5.7.3. Short-pitch corrugations	102
	5.7.4. Long-pitch corrugations	102
	5.7.5. Longitudinal vertical cracking	104
	5.7.6. Lateral wear	104
	5.7.7. Rolling surface disintegration	104

	5.7.8.	Shelling of the running surface	104
	5.7.9.	Gauge-corner shelling	104
5.8.	Permissible rail wear		105
	5.8.1.	Vertical wear	105
	5.8.2.	Lateral wear	105
5.9.	Optimum lifetime of a rail		106
5.10.	Fishplates		107
5.11.	The continuous-welded rail		109
	5.11.1. The continuous-welding technique		109
	5.11.2. Mechanical behaviour of the continuous-welded rail		109
		5.11.2.1. Assumptions	109
		5.11.2.2. Calculation procedure	110
		5.11.2.3. Force distribution along continuous-welded rail	111
		5.11.2.4. Length changes in the expansion zone	112
		5.11.2.5. Rail welding	112
		5.11.2.5.1. Flash-butt welding	112
		5.11.2.5.2. Thermit welding	113
		5.11.2.6. Destressing of a continuous-welded rail	113
	5.11.3. Expansion devices		114
	5.11.4. Advantages of the continuous-welded rail		115

6. Sleepers - Fastenings   116

6.1.	The various types of sleepers		116
6.2.	Steel sleepers		117
	6.2.1.	Form and properties	117
	6.2.2.	Manufacturing, size and weight	117
	6.2.3.	Advantages and disadvantages	119
	6.2.4.	Lifetime	119
6.3.	Timber sleepers		119
	6.3.1.	Form and properties	119
	6.3.2.	Dimensions	120
	6.3.3.	Advantages and disadvantages	121
	6.3.4.	Lifetime	121
	6.3.5.	Timber sleeper deformability	122
6.4.	Concrete sleepers		122
	6.4.1.	Inherent weaknesses of concrete sleepers	122
	6.4.2.	The two types of concrete sleepers	123
6.5.	The twin-block reinforced-concrete sleeper		124
	6.5.1.	Geometry and mechanical strength	124
	6.5.2.	Advantages and disadvantages	125
	6.5.3.	Lifetime	125
	6.5.4.	Deformability of the twin-block sleeper	126

6.6.	The prestressed-concrete monoblock sleeper	126
	6.6.1. Geometry and mechanical strength	126
	6.6.2. Advantages and disadvantages	127
	6.6.3. Lifetime	129
	6.6.4. Deformability of monoblock sleepers	129
6.7.	Stresses developing under the sleepers	129
6.8.	Track on concrete slab	130
	6.8.1. The two forms of slab track	130
	6.8.2. Problems created on a slab track	131
	6.8.3. Advantages and disadvantages of slab track	131
	6.8.4. Geometrical characteristics of a slab track	132
6.9.	Fastenings	132
	6.9.1. Functional characteristics	132
	6.9.2. Types of fastenings	133
	6.9.2.1. Rigid fastenings	133
	6.9.2.2. Elastic fastenings	133
	6.9.2.3. Types of elastic fastenings	135
	6.9.2.4. Operating principles of elastic fastenings	135
	6.9.3. Force developed in rigid and in elastic fastenings	137
	6.9.4. Rail creep and anti-creep anchors	138
6.10.	Resilient pads	138
	6.10.1. Rail seating and baseplate pads	138
	6.10.2. Functions and properties of pads	139
	6.10.3. Dimensions and materials	140
6.11.	Numerical application for the dimensioning of the various track components	140

7. Ballast   142

7.1.	Functions of the ballast and subballast	142
7.2.	Geometrical characteristics of ballast	143
7.3.	Mechanical behaviour of ballast	144
	7.3.1. Stress-strain relationship	144
	7.3.2. Fatigue behaviour	144
7.4.	Ballast hardness	145
	7.4.1. The Deval test	145
	7.4.2. The Los Angeles test	146
	7.4.3. Required ballast strength and hardness	147
7.5.	Ballast dimensioning	148
	7.5.1. The general dimensioning equation	148
	7.5.2. Numerical application	149
7.6.	Ballast cross-sections	149
7.7.	Laying the track	153
	7.7.1. Mechanical equipment	153

Contents

		7.7.2. Sequence of construction of the various track works	154

8. Transverse Effects - Derailment — 156

- 8.1. Transverse effects — 156
- 8.2. Transverse track forces — 156
  - 8.2.1. Transverse static force — 156
  - 8.2.2. Transverse dynamic force — 157
- 8.3. Transverse track resistance — 157
- 8.4. Influence of ballast characteristics on transverse track resistance — 158
  - 8.4.1. Influence of the geometrical characteristics of the ballast cross-section — 158
  - 8.4.2. Influence of the granulometric composition of ballast — 160
  - 8.4.3. Influence of the degree of ballast compacting — 160
- 8.5. Influence of sleeper type and characteristics on transverse track resistance — 162
- 8.6. Additional measures and special equipment used to increase transverse resistance — 163
- 8.7. Derailment — 164
  - 8.7.1. Derailment caused by track shifting — 164
  - 8.7.2. Derailment caused by wheel rebounding on the rail — 165
  - 8.7.3. Derailment caused by overturning of the vehicle — 166
  - 8.7.4. Derailment safety factor - numerical application — 166

9. Track Layout — 168

- 9.1. Rail vehicle running on curve and on transition arc — 168
  - 9.1.1. Effects during movement on curve — 168
  - 9.1.2. Transition arc - cubic parabola - clothoid — 168
- 9.2. Theoretical and actual values of cant - permissible values of transverse acceleration — 169
  - 9.2.1. Theoretical value of cant for complete compensation of centrifugal forces — 169
  - 9.2.2. Normal or actual value of cant — 170
  - 9.2.3. Permissible values of transverse acceleration — 171
  - 9.2.4. Cant deficiency variation in time — 172
- 9.3. Limit values of cant and acceleration — 172
- 9.4. Calculation of transition curve — 172
- 9.5. Calculation of the circular arc — 175
- 9.6. Case of consecutive same sense and antisense curves — 176
- 9.7. Superelevation ramp — 177
- 9.8. Combining maximum and minimum speeds — 178

9.9.	Correlation of the speed on a curve and of its curvature radius	179
9.10.	Gradients	180
9.11.	Vertical transition curves	180
9.12.	Layout design using tables	182
9.13.	Layout design using computer methods	182
9.14.	Construction of a new railway line	183
	9.14.1. Feasibility studies	183
	9.14.2. Preliminary design	184
	9.14.3. Outline design	184
	9.14.4. Final design	185
	9.14.5. Staking of the track layout	185

10. Switches and Crossings — 187

10.1.	Purpose of switches and crossings and their functions	187
10.2.	Arrangement and components of a turnout	188
10.3.	Various forms of turnouts	190
10.4.	Running speed on turnouts	192
10.5.	Derailment protection on switches and crossings	193
10.6.	Switches and crossings for high speeds	193
10.7.	Sleeper and track layout in turnouts and crossings	194
10.8.	Manual and automatic switch operation	194

11. Track Maintenance — 197

11.1.	Parameters influencing track maintenance	197
11.2.	Definitions and parameters associated with track defects	198
11.3.	Track defects	199
	11.3.1. Longitudinal defect	199
	11.3.2. Transverse defect	200
	11.3.3. Horizontal defect	200
	11.3.4. Gauge deviations	200
	11.3.5. Local distortion	201
11.4.	Track defect recording methods	202
11.5.	Limit values of track defects	203
11.6.	Progress of track defects	204
	11.6.1. Longitudinal defects	204
	11.6.1.1. Mean settlement of track	204
	11.6.1.2. Standard deviation of the longitudinal defect	205
	11.6.1.3. Interval between maintenance sessions	205
	11.6.2. Transverse defect	206
	11.6.3. Horizontal defects	206

11.6.4. Gauge deviations	207
11.6.5. Local distortion	207
11.7. Mechanical equipment for maintenance works	207
11.8. Scheduling maintenance operations	209
11.9. Technical considerations for track maintenance works	210
11.10. Weed control	212

12. Train Dynamics — 213

12.1. Train traction	213
12.2. Forces acting during train motion	213
12.2.1. The various kinds of forces	213
12.2.2. Running resistances	214
12.2.3. Empirical relations for the running resistances	216
12.2.3.1. Formulas of the French Railways	216
12.2.3.1.1. Diesel or electric locomotives	216
12.2.3.1.2. Pulled rolling stock	216
12.2.3.1.3. Electric passenger vehicles	217
12.2.3.2. Formula of the American Railways	217
12.2.3.3. Formula of the German Railways	218
12.2.3.4. Formulas for broad and narrow gauge railways	219
12.2.4. Resistances developed when running in a tunnel	219
12.2.4.1. Pressure problems	219
12.2.4.2. Increased aerodynamic resistance	221
12.2.4.3. Trains crossing in tunnels	223
12.2.4.4. Tunnel cross-section requirements at high speeds	223
12.2.5. Comparative running resistances between railways and road vehicles	224
12.2.6. Resistance due to track curves	224
12.2.7. Resistance caused by gravity	225
12.2.8. Inertial (acceleration) resistance	225
12.3. Specific traction force or starting force of a train	226
12.4. Adhesion forces	227
12.5. Required train power	229
12.6. Train acceleration and deceleration	230
12.7. Train braking	231
12.7.1. Braking systems	231
12.7.2. Braking distance	232
12.8. Line scheduling	235
12.9. Components of a pulled rail vehicle	236
12.9.1. Wheels	236
12.9.2. Axles	237

12.9.3. Bogies	238
12.9.4. Springs	239
12.9.5. Couplings and buffers	240
12.9.5.1. Springs in couplings	240
12.9.5.2. Springs in buffers	241
12.9.5.3. Couplings	241
12.9.5.4. Buffers	241
12.10. Dynamic stability of a rail vehicle	241
12.11. Vehicles with body tilting on curves	243

13. Diesel and Electric Traction   245

13.1. The various traction systems	245
13.2. Steam traction	245
13.2.1. Operating principle of the steam engine	245
13.2.2. Main parts of a steam locomotive	246
13.2.3. Disadvantages and abandonment of the steam locomotive	246
13.3. From steam traction to diesel traction and electric traction	247
13.3.1. From steam traction to diesel traction	247
13.3.2. From diesel traction to electric traction	247
13.3.3. Gas turbine locomotives	247
13.4. Diesel traction	247
13.4.1. Operating principle of the diesel engine	247
13.4.2. Transmission systems	249
13.4.3. Requirements from a diesel locomotive	249
13.4.4. Advantages - disadvantages of diesel traction	249
13.5. Electric traction	250
13.5.1. Power supply subsystem	250
13.5.2. Traction subsystem	251
13.5.3. Requirements of the two discrete subsystems	251
13.6. Electric traction systems	251
13.6.1. Direct current traction	251
13.6.2. Alternating current traction	252
13.6.2.1. AC traction at 15,000 V, 16 2/3 Hz	252
13.6.2.2. AC traction at 25,000 V, 50 Hz	252
13.6.3. Advantages and disadvantages of electric traction compared to diesel	254
13.7. Feasibility analysis of electric traction	256
13.7.1. Feasibility analysis parameters and procedure	256
13.7.2. Criterion for selection of the lines to be electrified	258
13.8. Transmission line	258
13.8.1. Parts and components of the transmission line	258

13.8.2.	Calculation of the transmission line cross-section	260
13.8.3.	Transmission line suspension	260
13.8.4.	Track configuration	261

13.8.5. Power transmission by conductor rail   261
13.9. Overhead line supporting poles   262
    13.9.1. Pole material   262
    13.9.2. Pole spacing   262
    13.9.3. Pole foundation   263
13.10. Substations   264
    13.10.1. Substations feeding direct current systems   264
    13.10.2. Substations feeding alternating current systems   264
    13.10.3. From thyristor to GTO technology   265
    13.10.4. Central remote control   266
    13.10.5. Interference of electrical traction on telecommunication systems   266
13.11. Synchronous and asynchronous motors   266
13.12. Electric locomotive maintenance - depot   267

List of References   269
Index   282

Preface

by

Ph. Roumeguère

Deputy Director General of the French Railways (SNCF),
Professor at the Ecole Nationale des Ponts et Chaussées,
Member of the Permanent Way Institution, of IRSE and of AREA

I think that the book Railway Engineering of V. Profillidis sums up his two principal characteristics: I had first met him in the beginning of the 80s as a graduate student at the Ecole Nationale des Ponts et Chaussées, where his Ph.D. thesis on the mathematical modelling of the railway infrastructure was brilliant. Since then, I meet him regularly amongst a worldwide circle of economic engineers, where he enjoys the respect of his colleagues, including myself.

Here we have a book which, unlike other books, is neither purely technical nor solely on economics, but provides a global view for the readers who are principally in the engineering technical domain. Indeed, what would be the use of rail technology, if it did not service the needs of transportation?

An important asset of the book, which justifies its publication, are the plentiful and well situated references, supporting and enriching the various chapters. It will be profitable, for instance, to make the connection between chapter 5 - chapter 4, chapter 7 - chapter 3, to better guide the reader's reasoning.

While there are books of high mathematical level - very few in fact, countable on the fingers of one only hand - I think that this book by V. Profillidis will be useful to a great number of students and professionals, who, I believe, will find the book accessible and easy to understand; thus the book avoids the destiny of ending up unopened and adorning a bookshelf. To his credit, V. Profillidis has, for instance, simplified the analysis of track mechanics, thus avoiding complicated elaboration.

I would eagerly recommend this book, either for the general programme of civil engineering studies or for the preparation of specializations.

Last, but not least, given that I am no great specialist in the English language, I appreciated that V. Profillidis' book was transparent to me, easy to go through, which in fact was the intended objective.

Foreword

Railways are today in a period of restructuring. New high speed tracks are being constructed and old tracks renewed, high comfort rolling stock vehicles are being introduced, logistics and combined transport are being developed. Awareness of environmental issues, daily highway and airport congestion and search for greater safety have given railways a new role within the transportation system.

Meanwhile, methods of analysis have significantly evolved and the use of computer applications has changed old railway methods and technologies. Perhaps the only common feature between the railway of the year 2000 and the railway of the 1960s is that both move on tracks, but with a totally different technology and design.

Railway engineering is the engineering discipline dealing with railway technology and operation. This dual nature defines an interdisciplinary approach: although it is considered within the competence of the engineer, a good knowledge of economic and social science is also necessary.

Like all technological and economic activities, railways have run their phases of development (from the second half of the 19th century to about 1950) and crisis (from 1950 to the mid-1980s). In the first phase they monopolized land transportation, while in the second phase they suffered a great reduction of their transportation activity as a result of competition from automobiles and airplanes.

However, for some years railways are in a phase of a new development, particularly in the sectors of high speeds, commuter services, freight traffic, combined transport, logistics, magnetic levitation trains, etc. Although their share in the transport market is poor (7% for passenger traffic and 17% for freight traffic in Europe), railways can expect a better future.

This book is intended for the use of railway engineers, consulting engineers and students of engineering schools and aims to provide a concise and useful synopsis of railway technology and scientific analyses that they will need in their daily scientific work or during studies. Each chapter contains a concise

Railway Engineering

theoretical analysis of the phenomena studied and applications, charts and design of the specific railway component. In this way, both the requirement for a theoretical analysis of phenomena (without exaggerated emphasis) is met, and the need of the engineer for tables, nomograms, regulations, etc. is satisfied.

It is customary to divide Railway Engineering into three activity functions: Track, Traction, and Operation (Commercial and Technical). This book contains civil engineering aspects of the railways. Thus, it is principally devoted to track topics. Given that electrification is a component of the permanent way and that train dynamics is necessary for the analysis of various aspects of the track, they are also included in this book. However, transportation aspects and some operation aspects are similarly included in the book. Signalling and other train control functions are not covered in this book, but they will make the topic of a future publication.

Railways present great differences in their technologies. Something may be valid for one such technology, but not for another. To overcome this problem, Regulations of the International Union of Railways (UIC) have been used to the greatest extent possible. Whenever a specific technology or method is presented, the limits of its application are clearly emphasized.

The book is the result of fifteen years of scientific, professional and teaching endeavours on railway subjects. Writing in its final form began in 1993 and was completed in early 1995.

I would like to address my warmest thanks to Mr. Ph. Roumeguère, Deputy Director General of the French Railways, for the preface he wrote for the book and for the exchange of views which we had about certain matters. I address also my thanks to Mr. W. Steinmetz and Mr. C. Bonnett of the Railway Strategy Center at the University of London-Imperial College, for having contributed useful suggestions. And I also appreciated the comments of Mr. S. Athanassopoulos.

Typing of a difficult technical text was carried out by Ms. Helen Pipinika and drawings were executed by Ms. Nancy Parinta. I would like to express to them my special thanks for their diligence and patience.

However, since nothing is absolute and permanent as regards knowledge, I will welcome the views and comments of the readers.

V. A. Profillidis

1 Railways and Transport

1.1. Evolution of the railways

1.1.1. Historical outline

Since the dawn of human activity to this day, quick and safe transportation of people and goods has been a constant goal of every organized society. It is generally acknowledged, (38)*, (1), that the fundamental innovations in the development of transportation included the discovery of the wheel, the railway and the airplane. Railways in their present form made their appearance at the beginning of the 19th century in British mines. Their main characteristic is the guided movement of the wheel by the track through a metal-to-metal contact, conferring to the rail vehicle a single degree of freedom.

However, the forerunners of the railways of our time appeared much earlier than the 19th century. Movement of carriages or wagons on metal guides is illustrated in a 1550 gravure found in Basel, Switzerland, which shows transportation methods employed in the mines of Alsace. The guided movement of carriages in general was already known in Roman times, as witnessed by grooves carved in the stone pavement to facilitate and speed up the movement of carriages, (1).

On Mount Penteli near Athens, where the white marble of the Parthenon and other classical monuments originated, deep grooves in the rocky ground still bear testimony to the methods employed by ancient Greeks to move marble slabs to the construction sites. Furthermore, according to certain authors, (38), the guided movement of carriages was applied in Greek antiquity by laying wooden channels on dirt roads to guide carts. Two channels were adequate for the needs of the day to accommodate one carriage. When two carriages came face to face, the younger driver would make way for the older driver. It was suggested that in such an encounter, Oedipus refused to make way and killed the older cart

---

* Figures between parentheses denote references, the list of which is at the end of the book.

1

driver coming from the opposite direction, unaware that it was his father Laïus, (38).

1.1.2. The golden age of the railways

Railway development was decisively influenced by the industrial revolution, the introduction of steam and the extensive exploitation of coal and iron mines. The first railway lines began operating in most European countries around 1830 and most railway networks attained maximum density at the beginning of the 20th century. A factor contributing the massive growth of the railways was high speed (by the standards of the time), which enabled fast connections. Steam-powered engines had already achieved (in test runs) impressive performances: 100 km/h in 1835 in Britain, 144 km/h in 1890 in France, 213 km/h in 1903 in Germany. Although maximum operating speeds were much lower (1/2 to 2/3 of test speed), they contributed to the rapid growth of rail transportation.

The adoption of electric traction in the early 20th century permitted further development of the railways, while the development of signalling and centralized remote control before World War II gave railways in the 1950s their present form.

1.1.3. Railways and other competing transportation means

Times have changed, however, and what was impressive in the early 20th century, was soon becoming less and less satisfactory. Airplanes and private cars were already offering transportation alternatives at every scale. Given the pressure of competition, the railways had to modernize and improve, especially as regards speed, reduction of transportation costs and better organization and improvement of the services offered. Hence, we come to the era of high-speed trains (Photos 1.1., 1.2.) operating at 250-300 km/h (a speed of 515 km/h was attained by French Railways in 1990 in test runs), combined transport (combined rail-road transportation), high-volume transport for both passengers (commuter service) and freight (bulk loads), (14), (18), (25), (26).

Nevertheless, in parallel with the conventional railways (which are based on metal-to-metal contact), experimental development proceeded since the mid-1970s with techniques which, although using guided vehicle travel (like railways), avoid any contact between the moving vehicle and the bearing substructure. These are the aerotrain (Photo 1.3) and the magnetic levitation train, or maglev, (Photo 1.4), which, in test runs, have attained speeds of 422 km/h for the aerotrain in 1969 and 600 km/h for the maglev in 1991, (39) (see also below section 1.7).

Railways and Transport

Photo 1.1. The Japanese Shinkansen high-speed train

Photo 1.2. The French TGV high-speed train

Photo 1.3. The aerotrain

Photo 1.4. The magnetic levitation train

Railway Engineering

1.1.4. The evolution of railways' organization

The organization of railway enterprises began in the late 19th and early 20th centuries in the form of small private businesses. The strategic importance of the railways for the economy and the security of various countries and the deficits which had already begun to appear, led most governments between 1935-1960 to nationalize their railways. Therefore, after the 1950s, most railways became part of the state administration, entailing on the one hand the organized development of rail transportation on a national scale and on the other hand the inflexibility and reluctance to modernize as well as the accumulation of deficits (1960s-1980s period).

Developments in the transport market at the end of the 1980s (mainly the gradual liberalization of transport activities from the regulating framework under which they had been operating for three decades or more), compelled railways to show more flexibility in the organization of their transport services, reduce transport costs, adapt to new technology, exploit their comparative advantages, and modernize, to make them competitive in the transport market. Some countries, like Japan, Great Britain, Sweden etc., have recently begun a privatization procedure of their national railways, (24). On the transport market, no technology and innovation will have a reason to exist unless they are financially efficient and competitive, compared to services offered by other transport means (road vehicles, airplanes), (32), (34), (41).

1.2. Rail transport characteristics

The main characteristic of rail transport involves the capability to join several transport units into trains. Therefore, in freight transport, trains of 15,000 tons gross with multiple coupling are used daily in the United States (indeed a test run was conducted with a freight train of 50,000 tons gross). In passenger service, railways are likewise capable of carrying a large number of people. Shinkansen high-speed trains of the Japanese Railways have in one day moved 520,000 passengers between Tokyo and Osaka (a distance of 515 km).

Another characteristic of rail transport is its one degree of freedom, in comparison to automobiles which have two. The one degree of freedom makes door-to-door transportation impossible for rail, but favours large-scale use of automatic controls, computers and electronics. As a result, unit transportation capacity (commuter trains are often moving 60,000 passengers per hour and per direction) is greatly increased.

As indicated in the previous paragraph, rail transport is characterized by the guided movement of wheels on tracks through the metal-to-metal contact, which considerably reduces rolling resistance (to less than 3 kg per ton carried). Accordingly, for the same propulsion force, rail vehicles carry a much larger load than road vehicles. As a result, rail transport consumes half as much energy as road transport for the same traffic. The comparison becomes more definitive with airplanes, which consume 5-7 times more energy than railways, (31), (32).

Another advantage of rail transport is much lower environmental pollution. Electric trains cause no pollution, while diesel-powered trains generate 15 times less pollution than automobiles for the same traffic, (32).

People all over the world have become more sensitive about transport safety. For the same traffic, the risk of a fatality occurring is eight times greater in road than in rail transport, whereas the risk of an injury is 200 times more likely to occur on a roadway than on rail, (32). It is a really impressive performance of the railways.

Finally, land occupation is much less for rail transport than for other transportation means, and specifically three times less than for road transportation. For the purposes of comparison with airplanes, it is noteworthy to mention that the new high-speed Paris-Lyons line (a distance of 429 km), occupies as much space as the new Paris airport at Roissy.

1.3. Evolution of railway traffic

1.3.1. Participation of private and public transport modes in the organization of the transportation system

The decades after 1950 showed a considerable increase in the mobility of individuals. The number of journeys increased greatly, mainly as a result of:

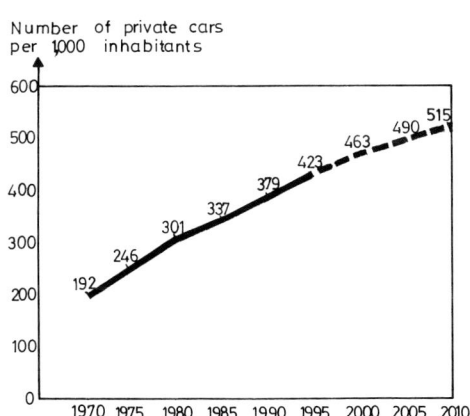

Fig. 1.1. Average private car ownership index in EU countries, (15)

(a) the population increase
(b) the increase in the standard of living, accompanied by an increase of the private car ownership index. The average value of this index for the European Union (EU) countries is currently (1995) 1 private car per 2.5 inhabitants, and it is estimated that in 2010 it will reach a value of 1 private car per 2 inhabitants (Fig. 1.1.). The private car ownership index is directly related to per capita national product, but not proportionally (Fig. 1.2), since it is influenced by the development of the various transport means for each country, by its geographical position, etc.

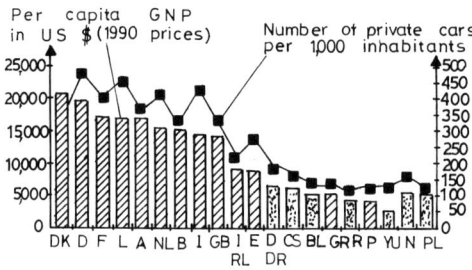

Fig. 1.2. Correlation of per capita GNP and private car ownership for various European countries (denoted by their initials), (15)

(c) the gradual reduction of the importance of borders, with the result that the mobility increase of land transport is greater in central European countries than that along the European periphery (Figs. 1.3, 1.4).

However, extreme differences are observed in car ownership among countries, as well as at its rate of evolution (Fig. 1.5).

The major part of this mobility increase was absorbed by private means transport (TPR) to the detriment of transport by public means (TPU). It was indeed found that TPR is an increasing function of the private car ownership index (PCO), while TPU is a decreasing function, given by the empirical formulas (1.1) and (1.2) (next page), (15):

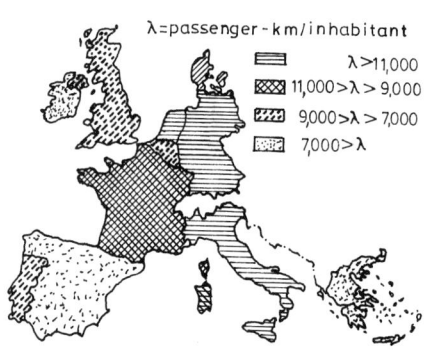

Fig. 1.3. Passenger land transport activity expressed in passenger-kilometres per inhabitant for EU countries, (28)

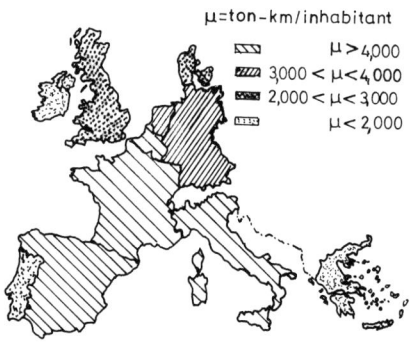

Fig. 1.4. Freight land transport activity expressed in ton-km per inhabitant for EU countries, (28)

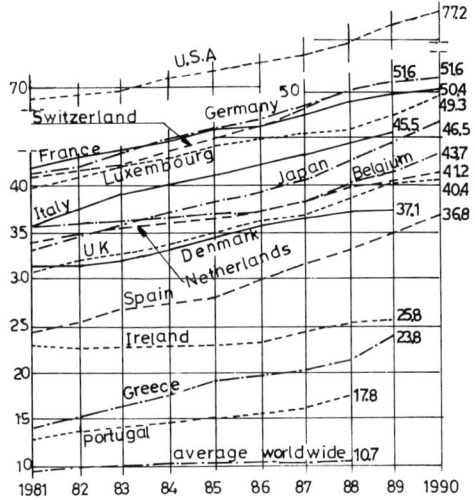

Fig. 1.5. Evolution of the private car ownership index (number of private cars per 100 inhabitants) in various countries in the 1980s, (10)

$$TPR = 5.7\left[1 - e^{\left(-\frac{PCO}{450}\right)}\right] \quad (1.1)$$

$$\frac{\text{Total Passeng. Trans.}}{TPU} = 1 + \frac{TPR}{TPU} \quad (1.2)$$

Consequently, public transport (including rail transport) did not benefit, to the degree it could have, from the mobility increase in recent decades.

The reasons for this stagnation in rail transport are analyzed in section 1.4.1. and are mainly focused on the following advantages of private road transport:

- door-to-door transport
- higher comfort
- greater speed and (accordingly) less travelling time
- flexibility
- low cost
- improvement of the image (as a result of systematic marketing and promotion efforts).

Only during the 1980s did the railways begin to provide solutions which could fulfil some of the above advantages of private road transport.

However, it is shown that the evolution of transport activity as a whole is at approximately the same rate as the evolution of the Gross National

Product, (32). Air transport rates are greater than GNP rates (almost double), whereas rail transport rates are slower, (24).

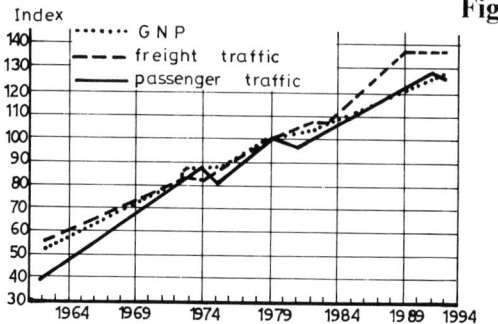

Fig. 1.6. Trends in the total passenger traffic (passenger-km), the total freight traffic (ton-km) in relation to the Gross National Product for the countries of the European Conference of Ministers of Transport (ECMT)* between 1962-1992, (32)

1.3.2. Share of rail in the transport market

In the Western European countries, the share of rail transport is low, with a medium value in 1992 of 7.2% for the passenger transport market and 17.0% for freight transport (See also tables 1.1 to 1.3 in the following section). Conversely, Eastern European countries (which for more than four decades had central economic planning), through state intervention, favoured a high participation of railways in the overall transportation activity. Thus the share of railways in the freight transport market for these countries was approximately 60-80% in the mid-1980s. However, after the liberalization of the economy of these countries, the share of the railways in the national transport market has been dramatically decreasing, over 50% to date in most cases between 1990-95.

It is therefore evident that the share of the railways in the national transport market depends mainly on the degree to which the railways meet the requirements of the market and society, as well as on the degree and orientation of state intervention. The gradual decrease of the latter leaves the adaptation of the railways to market requirements as the only critical parameter determining their position within the market.

1.3.3. Evolution of rail passenger traffic

With 1970 as a reference year for transport activity (in passenger-kilometres) for each transport mode, private automobiles during the period between 1970-1992 more than doubled their transport activity, buses increased by 50% and the railways by 40% (Figs. 1.7, 1.8), (20), (20a), (25), (42). Thus, the share of the railways in passenger transport in 1992 for the ECMT countries was limited to only 7.2% (Table 1.1).

---

* Austria, Belgium, Denmark, Finland, France, Germany, Greece, Iceland, Ireland, Italy, Luxembourg, Norway, Portugal, Spain, Sweden, Switzerland, Turkey, U.K., former Yugoslavia.

Railways and Transport

1.3.4. Evolution of rail freight traffic

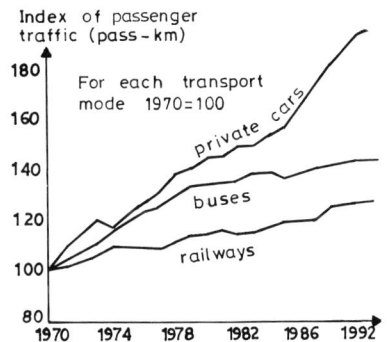

Fig. 1.7. Evolution of passenger traffic development index for various transport modes in ECMT countries, (20a)

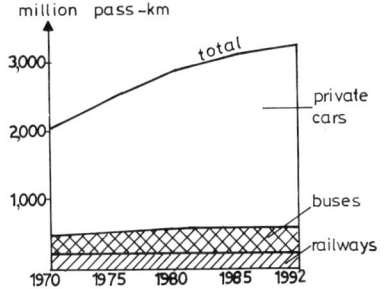

Fig. 1.8. Evolution of passenger traffic for various transport modes in ECMT countries, (20a)

Similarly for freight transport, using 1970 as the reference year for the activity (in ton-kilometres) for each transport mode, road carriers more than doubled their transport activity between 1970-1992, while the railways stagnated (Figs. 1.9, 1.10). Hence, the rail share with respect to the overall freight transport was limited to 17% (Table 1.2, next page). Table 1.3 illustrates the rail freight traffic for the various continents of the world.

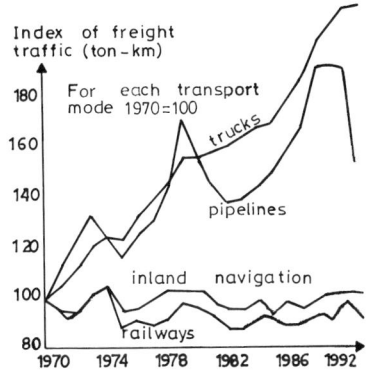

Fig. 1.9. Evolution of freight traffic development index for various transport modes, (20a)

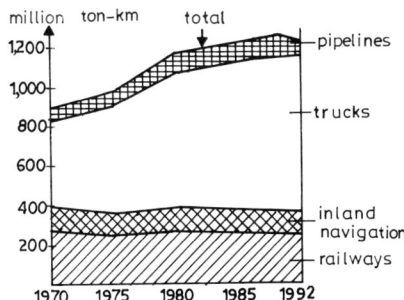

Fig. 1.10. Evolution of freight traffic for various transport modes in ECMT countries, (20a)

Table 1.1.
Evolution of the share (%) of various transport modes in passenger traffic in ECMT countries, (20a)

	1970	1980	1986	1992
Private cars	77.3	79.7	81.4	83.4
Railways	10.7	8.7	8.1	7.2
Buses	12	11.6	10.5	9.4

Table 1.2.
Evolution of the share (%) of various transport modes in freight transportation in ECMT countries, (20a)

	1970	1980	1985	1990	1992
Trucks	55.2	65.9	69.3	74.0	75.1
Railways	31.3	23.2	21.2	17.4	17.0
Inland Navig.	13.5	10.9	9.5	8.6	7.9

Table 1.3.
Rail freight traffic in the various continents of the world for the year 1990

Geographic area	Freight traffic (in billion ton-kms)
Europe	515
Asia	1,500
(China)	(1,100)
Former Soviet Union	3,700
America	2,000
(USA)	(1,500)
Africa	120
Australia	80

1.3.5. Railway networks and transportation activity in European countries

Table 1.4 gives characteristics of European railway networks:
- length of railway lines
- length of electrified railway lines
- passenger activity (in passenger and passenger-kilometres) in 1992
- freight activity (in tons and ton-kilometres) in 1992
- staff numbers.

Another important fact is productivity, that is the transport output per employee. Table 1.5 illustrates for the railways of the principal European countries the productivity in 1991 (thousands of traffic units per employee), (32).

Table 1.5.
Railway productivity in several European countries, (32)

Country	Railway Productivity in 1991 (in thousands of traffic units per employee)	Country	Railway Productivity in 1991 (in thousands of traffic units per employee)
Austria	352	Italy	383
Belgium	405	Luxembourg	281
Denmark	438	Netherlands	587
Finland	597	Poland	359
France	598	Portugal	354
Germany	350	Spain	562
Greece	213	Sweden	1,034
Hungary	167	Switzerland	587
Ireland	358	United Kingdom	370

Table 1.4
Railway networks and traffic in European countries in 1992, (20)

Country	Railway Network	Length of railway lines (km)	Length of electrified railway lines (km)	Passenger traffic		Freight traffic		Staff numbers
				Passengers (millions)	Passenger-kms (billions)	Tons (millions)	Ton-kms (billions)	
Austria	OBB	5,605	3,246	174.935	9.561	65.423	12.609	65,793
Belgium	SNCB/NMBS	3,432	2,291	145.006	6.798	77.534	9.657	43,889
Bulgaria	BDZ	4,294	2,650	75.909	5.393	32.261	7.758	56,585
Czechoslovakia	CSD	13,099	3,965	395.928	16.898	170.471	44.187	183,226
Denmark	DSB	2,306	280	142.872	4.600	7.721	1.747	20,076
Finland	VR	5,874	1,664	45.140	3.057	31.206	7.431	18,945
France	SNCF	32,731	12,986	820.357	62.647	132.744	48.193	198,078
Germany	DB	26,779	12,149	1,098.327	46.407	249.607	55.064	226,854
	DR	14,054	4,332	330.865	9.839	85.136	13.625	210,175
Greece	OSE	2,484	–	11.819	2.004	3.397	564	12,458
Hungary	MAV	7,727	2,306	138.659	6.820	51.541	9.576	99,635
Ireland	CIE	1,944	37	25.837	1.226	3.333	633	5,163
Italy	FS	16,112	9,936	440.000	48.361	70.760	22.548	167,566
Luxembourg	CFL	275	220	10.371	220	16.817	672	3,471
Netherlands	NS	2,753	1,987	332.500	15.350	17.124	2.764	28,348
Norway	NSB	4,027	2,426	35.800	2.256	16.038	2.161	12,450
Poland	PKP	25,254	11,496	549.302	32.571	201.702	57.774	275,113
Portugal	CP	3,062	461	224.621	5.694	8.169	1.907	20,269
Romania	CFR	11,430	3,782	323.758	24.269	110.334	24.169	189,565
Spain	RENFE	13,041	6,894	358.610	16.350	28.272	11.372	47,867
Sweden	BV (SJ)	9,846	7,268	76.600	5.234	51.577	19.235	24,943
Switzerland	SBB/CFF	2,985	2,969	267.537	11.819	46.097	7.513	38,271
U. K.	BR	16,528	4,910	744.742	31.718	122.356	15.508	137,729

Railway Engineering

However, Figs. 1.11 and 1.12 illustrate the evolution of both passenger and freight rail traffic in Europe from 1980 to 1993. Traffic has remained almost stagnant in countries of the European Union and the EFTA, whereas in the former socialist countries of Central Europe it has fallen dramatically after the beginning of liberalization of their economies (1989).

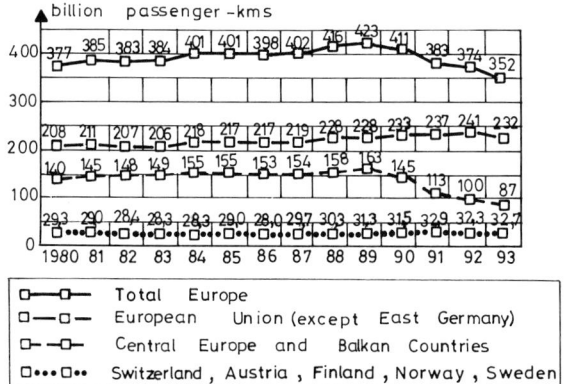

Fig. 1.11. Evolution of rail passenger traffic in Europe, (20)

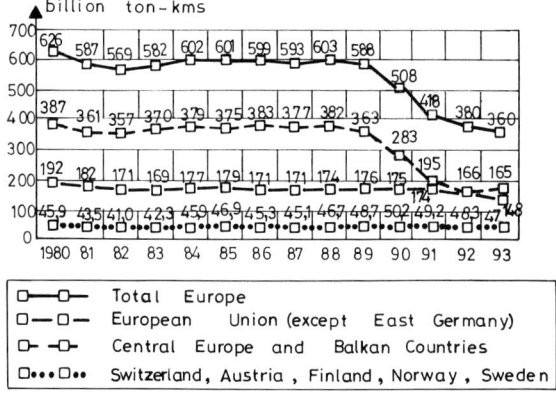

Fig. 1.12. Evolution of rail freight traffic in Europe, (20)

1.3.6. Forecasts for the evolution of the transport market

As previously stated in section 1.3.1, the evolution of transport activity is very closely related to the evolution of the development of the economy and specifically to the rate of the Gross National Product (GNP). ECMT countries had an average annual increase in their GNP of 2.3% between 1973 - 1990.

Railways and Transport

The economic situation in the mid-1990s was characterized by stagnation and the rate of increase of the GNP for the period between 1995-2000 is estimated at approximately 1.5% for the ECMT countries and 2.5% for the European Union countries.

The evolution of transport activity can be forecasted with the help of the appropriate models, which take into account parameters such as GNP, population, mobility, etc. Various scenarios are studied and usually include a pessimistic, a realistic (medium) and an optimistic one. As a basic scenario the realistic one is used, (41).

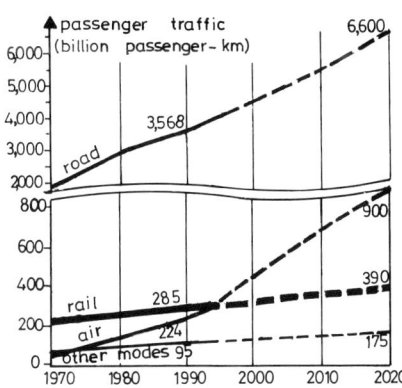

Figure 1.13 shows ECMT forecasts for passenger traffic. Total passenger traffic will increase by 2.4% per year until 2010 and begin decreasing thereafter. Railways will have an annual increase of 1.0%, road transport 2.2%, whereas air transport will have the highest rate, at approximately 6%, according to the ECMT forecasts, (32).

Fig. 1.13. Forecasts for the evolution of passenger traffic in Europe, (32)

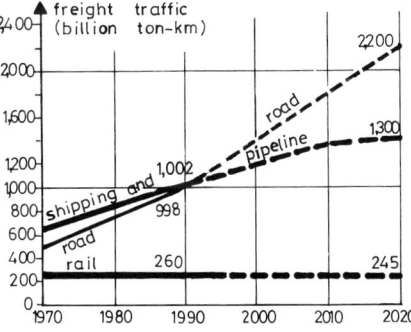

Figure 1.14 shows ECMT forecasts for freight traffic. Total freight traffic will increase by 2.0% per year until 2010. Rail traffic will decrease by 0.3% per year, road traffic will increase by 3.0% per year, whereas shipping and pipelines are anticipated to have a medium annual increase of around 1.5% according to the ECMT forecasts, (32).

Fig. 1.14. Forecasts for the evolution of freight traffic in Europe, (32)

1.4. The dual nature of the railways: business and technology

1.4.1. Weaknesses inherited by the railways from the past

The foregoing analysis of the evolution of the railways clearly shows that the existence of a railway line, constructed 100-150 years ago, can in no case be a sufficient justification to continue operation of the particular line. Railways should search for their comparative advantages, which they should develop by the necessary technological modernization. On the other hand, they should operate as enterprises governed by the same rules of competition applied in other businesses, relinquishing the umbrella of state protectionism sheltering them for decades.

Railways, however, inherit serious handicaps as a result of decades of state protectionism, (11), (12), (34):

- administration and organization inflexibility. For decades, railway management was dealing only with current affairs. Important matters were conducted by the supervising ministry, often based on political criteria,
- accumulation of personnel in routine tasks and staff shortages for administration, organization and technological upgrading positions,
- high transportation cost, often the result of obsolete operating methods,
- rolling stock often difficult to operate, offering services of a level which in many instances does not meet current requirements,
- railway infrastructure maintenance expenses mostly borne by the railway enterprise, as contrasted to road carriers which contribute only a small fraction of the road network maintenance cost, and to air carriers which contribute a very small fraction of airport maintenance cost.

  An important institutional evolution on this matter was EEC Directive 440/1991 stating that the railways of the EEC countries should separate their accounts for infrastructure expenses from those concerning operation. Infrastructure accounts for all transport modes will be the responsibility of the state, (24),
- obsolete infrastructure, often as a result of the absence of serious investment for many decades. The same applies to rolling stock,
- obligation to operate lines with little transport activity, which the railway enterprise, were it operating by private enterprise criteria, would not have kept in operation,
- As a result of all the above, railway deficits have been steadily increasing for more than three decades.

1.4.2. Comparative advantages of the railways

The aforementioned compendium of disadvantages risks giving the impression that the railways have nothing but problems (which to a large degree are of the making of others). However, the contribution of railways to the development of both transport and the economy is by no means negligible, since railways, (14), (27):

- provide an integrated system of services for both passenger and freight transport with programmed schedules regardless of day and season,
- provide reduced fares for large segments of society (e.g. students, the elderly, etc.) who can thus travel easier,

- pollute the environment minimally in contrast to other transport means,
- contribute decisively to relieve congestion in peak travel periods in central thoroughfares because of huge transport capacity,
- consume much less energy for the same traffic than by any other transport mode.

1.4.3. Development strategy and measures to restructure the railways

The transport sector in Europe and worldwide is presently oriented to a gradual deregulation and liberalization, with emphasis on competition between the various transport modes. The government-owners of the railways are under the obligation to ensure a real autonomy for the railways, to gradually reduce subsidies to railway enterprises (used to cover deficits), to institute a regime of transparency in railway operations and to create a frame-work in which other rail companies can use the railway infrastructure and enter the rail transport market. Within such a frame-work, the railways should aim at, (42), (24), (11), (34):

- greater flexibility in the organization and development of operational criteria for the various alternatives, e.g. investment,
- personnel allocation on the basis of the requirements of the particular transportation task and staffing of the various departments by specialized personnel. It is not to exclude, particularly for management and specialized tasks, the use of high quality specialists from other sectors,
- trying to drastically reduce costs in order to make rail services more competitive in the transport market. The reduction of cost may come from the application of informatics and new technologies in addition to the rationalization and inevitable reduction of the current personnel levels,
- systematic maintenance and renovation of the rolling stock and infrastructure enabling the railways to meet the requirements of their clients,
- separation of infrastructure maintenance expenses from other expenditures. Maintenance expenses may be the responsibility of the state (as are the road network and the airports) or of a company clearly different from rail operation,
- infrastructure modernization with important investment (for the most part this can be covered by the state, the EU, the World Bank, etc.). It should be stressed here that modernization does not refer to any particular project, but to those that will enable railways to competitively coexist with other transport means. For the more attractive projects, financing can come from the private sector also, as was the case of the Channel Tunnel Project,
- clear definition of public service obligations, being understood as those which, if the only consideration of the enterprise were business profit, would not have been undertaken to the same extent or degree (e.g. exploitation of lines with small traffic). The agency enforcing a mandatory public service (e.g. the Ministry of Education for reduced pupil-student fares) should refund lost income to the railway enterprise,

- adequate compensation of the railways for not polluting the environment and not causing traffic congestion. A quantitative and financial evaluation of the effects of the various transport modes on the environment is already available, (43). The prevailing view is to subsidize railways with an amount corresponding to that which would have to be expended to combat the pollution and traffic congestion which would have been caused had operation of the railway been discontinued,
- gradual reduction of deficits.

Fulfilment of the above conditions is a necessary requirement for the survival and growth of the railways within the transport market.

1.5. High-speed trains

1.5.1. The application of high-speeds in railways

High-speed trains (with V > 200 km/h) were the railway response to the transport market requirement for reduced travel times. High speeds were pioneered by two railway networks:

- the Japanese network, with the 1964 commissioning of the "Shinkansen" high-speed rail link between Tokyo and Osaka, with a top speed of 210 km/h,
- the French Railways, by inaugurating in 1981 the "TGV" high-speed train between Paris and Lyons, with a top speed of 260 km/h, increased to 270 km/h in 1983 and to 300 km/h in 1989.

Both lines were built on heavily travelled routes showing signs of saturation. Faced with improving the existing infrastructure or building a new high-speed line, the latter was opted. High-speed lines were constructed in the 1980s in West Germany (Hannover-Würzburg and Mannheim-Stuttgart), in Italy (Rome-Florence), in Spain (Madrid-Sevilla), in France ("TGV Atlantique" line Paris-Bordeaux) and recently the Paris-London line (through the Channel Tunnel).

Two approaches to high speeds can be distinguished:

- In the first, only passenger trains run on high-speed lines, with low loads per axle, very small tolerances on track defects, and large gradients (up to 35 ‰). This approach was implemented in the Paris-Lyons line and presupposes a high passenger train traffic to make the construction and operation of the new line cost-efficient, (45)
- In the second, the new high-speed lines are run by both passenger and freight trains, the coexistence of which entails higher maintenance costs and requires lower values of longitudinal gradient, (23), (40). Most high-speed lines are currently designed for mixed traffic (both passenger and freight trains).

1.5.2. Impact of high speeds on the reduction of rail travel times

The reduction of travel time was a constant goal of the railways, as can be seen from Table 1.6. Only with high-speed trains, however, were the railways able to achieve on 400 ÷ 700 km routes travel times equal to or better than air travel.

Table 1.6.
Train travel time reduction on certain routes

	1937	1950	1960	1980	1983 (TGV line, length 427 km)	1987
Paris-Lyons (511 km)	5h15min	5h05min	4h00min	3h50min	2h00min	1h50min

	1963	1965 (Shinkansen)
Tokyo-Osaka (515 km)	5h30min	3h10min

Indeed, high-speed railways capitalize on their advantage to reach city centres and thus make travel times from the centre of a city to the centre of another far shorter than for automobiles and even, in many cases, shorter than for airplanes (Table 1.7).

Table 1.7.
Comparison of travel times from the centre of a city to the centre of another for trains, airplanes, and automobiles (case of the Paris-Lyons route), (23)

TGV travel time	TGV travel time + access time to railway station	Airplane [1]	Automobile [2] (on highway at a top speed of 120 km/h)
1h50min	2h30min	2h30min ÷ 3h	5h

[1] The time indicated is the sum of the flight time, travel time from the city centre to the airport, and check in time and retrieval time.

[2] The time indicated is the time from the centre of a city to the centre of another, i.e. it takes into account the time necessary (about 30 min) for a car to reach the highway from the city centre.

1.5.3. High speeds and the increase of traffic

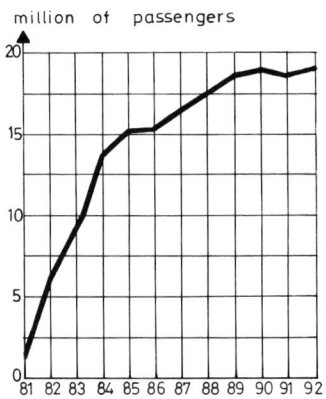

Fig. 1.15. Evolution of traffic on the Paris-Lyons TGV, (45)

Another result of high speeds was the increase of traffic (Fig. 1.15), either derived from air and automobile transport (diverted demand) or as totally new activity (generated demand).

High speeds therefore attract back to the railways part of the passenger traffic lost in the past. For this purpose, however, a speed increase is not enough, station accessibility should also be improved through efficient bus or metro systems. In many instances, connection of railway stations serving high speed trains to the airports can contribute to an efficient (from time and cost point of vue) air-rail trip.

1.5.4. The European high-speed rail network

The success of high-speed trains, which are especially suitable for distances on the scale of Central and Western Europe, led the European Commission to prepare a plan to implement a high-speed European rail network (Fig. 1.16), (21). This network will include both new lines (with design speeds exceeding 250 km/h) and existing lines the layout of which will be improved to accommodate speeds of 200 km/h. The European high-speed network is expected to be realized by 2010. Figure 1.17 shows the effects of the realization of the European high-speed network on travel time reduction for some routes.

1.5.5. Technical features of high-speed railway lines

Table 1.8 shows the technical characteristics of high-speed rail lines. Important differences as regards gradients and electric traction systems are observed. Implementation of the European high-speed rail network thus requires that many technical characteristics, which now show differences as a result of industrial and technological production particularities in each country, must be standardized.

However, railway authorities aim at even higher running speeds, (5). Thus, French Railways schedule a maximum running speed of 360 km/h for the year 2000 and of 400 km/h around 2010, whereas the German Railways schedule a running speed of 330 km/h for the year 1998.

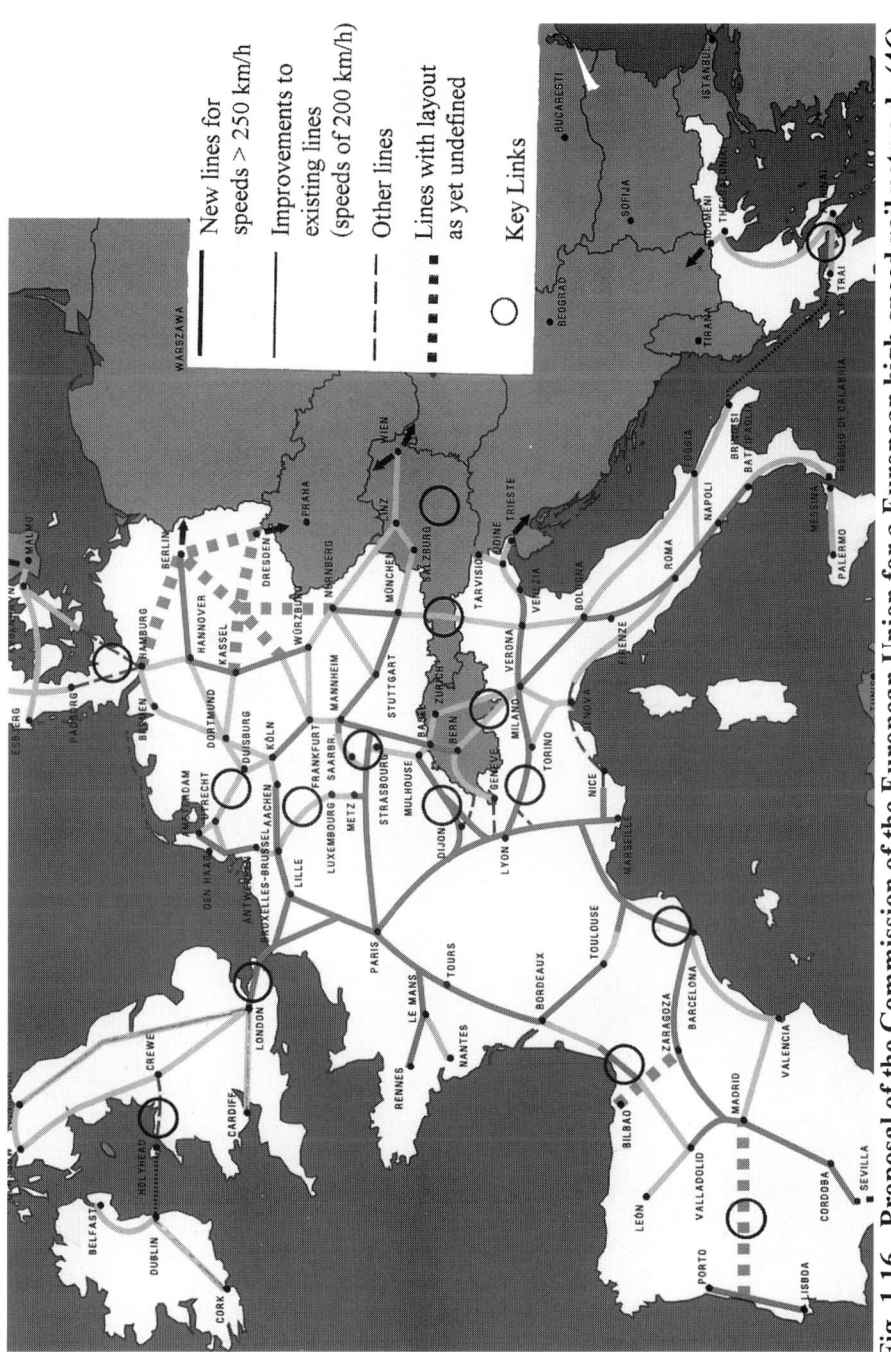

Fig. 1.16. Proposal of the Commission of the European Union for a European high-speed rail network, (46).

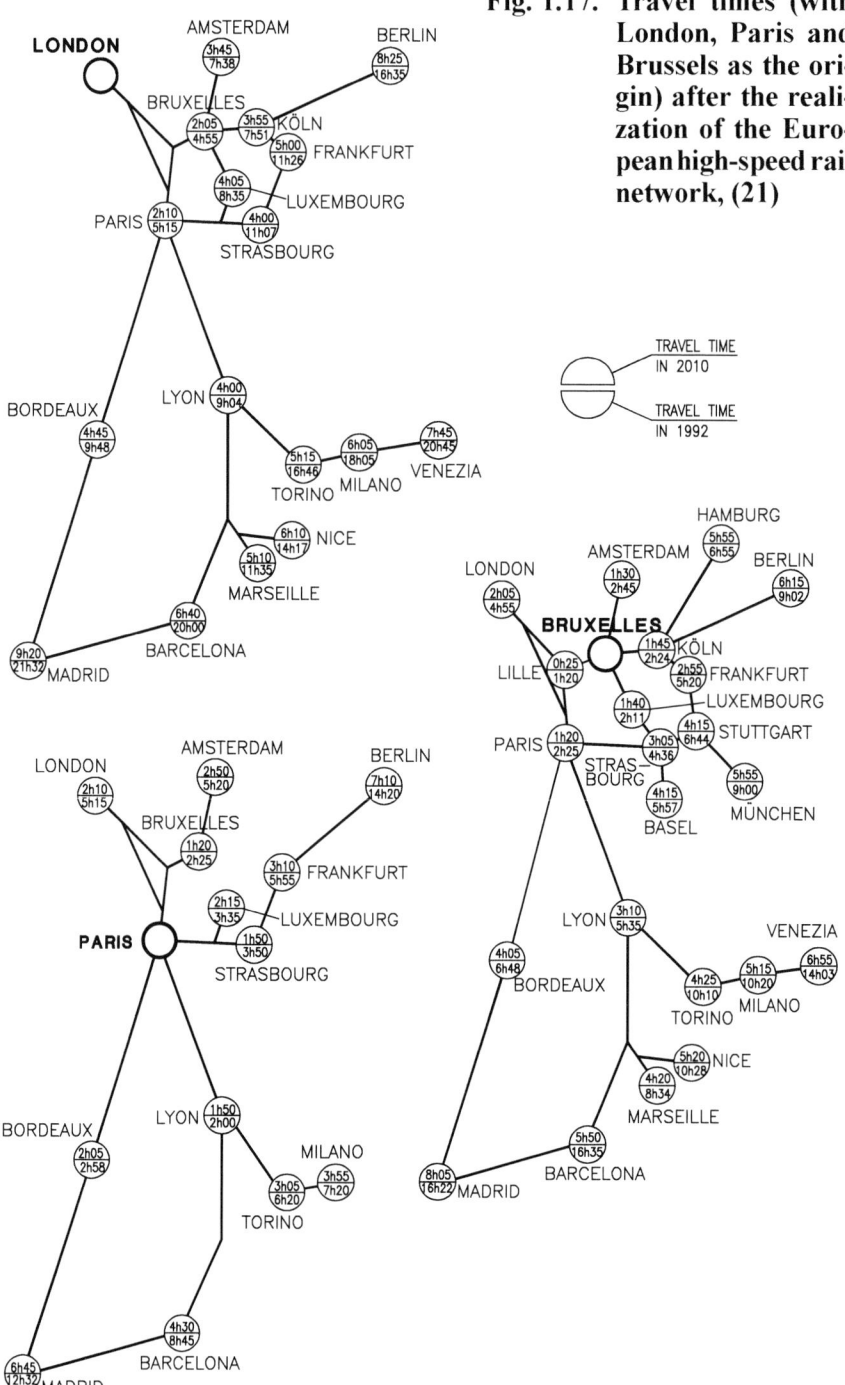

Fig. 1.17. Travel times (with London, Paris and Brussels as the origin) after the realization of the European high-speed rail network, (21)

Railways and Transport

Table 1.8.
Technical characteristics of high-speed rail lines, (23), (40), (44)

Country	France	Germany	Italy	France	Spain
Line	Paris-Lyons (427 km)	Hannover-Würzburg (327 km)	Rome-Florence (260 km)	Paris-Bordeaux (260 km)	Madrid-Barcelona (522 km)
Design speed V_{max} (km/h)	300	250	250	300	300
Radius of curvature R_{min} (m)	4,000	7,000	3,000	4,000	4,000
Maximum longitudinal gradient (‰)	35	12.5	8	25	30
Traction power supply	25 KV 50 Hz	15 KV $16\frac{2}{3}$ Hz	3 KV	25 KV 50 Hz	25 KV 50 Hz

1.6. The Channel Tunnel Project

1.6.1. Project description

After more than a century of efforts, the Governments of the United Kingdom and France decided in 1986 on a permanent railway link between the two countries, that should be realized entirely by private financing. For this purpose the Eurotunnel Consortium was created with responsibilities to construct the tunnel and operate it for 55 years.

The project of a total length of 50 km consists of two rail tunnels (one per direction) with an internal diameter of 7.6 m plus a third tunnel (of an internal diameter of 4.8 m), for maintenance purposes, emergency incidents etc. (Fig. 1.18). The principal tunnels are

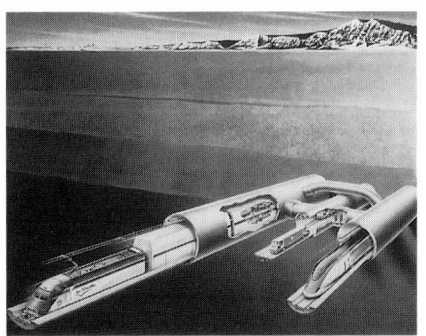

Fig. 1.18. The Channel Tunnel

Railway Engineering

connected to the auxiliary one at 375 m intervals. The rail level is situated 25-40 m below the seabed level.

The entire construction cost, that was initially underestimated and changed many times, is allocated as follows:
- 50% for the tunnel construction
- 10% for the rolling stock
- 40% for tracks, signalling, electrification, etc.

1.6.2. Travel times and forecasted demand

Full operation through the Tunnel began in autumn of 1994. Four types of services are provided:
- high speed trains, (named Eurostar) with a running speed in the tunnel of 160 km/h, joining London to Paris in 3h and London to Brussels in 3h15. Eurostar trains have a capacity of 794 passengers (584 in second class and 210 in first class)
- conventional trains, night trains, freight trains (transporting containers, new cars, etc.), with a usual speed of $100 \div 120$ km/h
- shuttle passenger trains (named "Le Shuttle"), transporting cars (in two floors) and trucks and buses (in one floor). Passengers remain in their seats and maximum speeds are 140 km/h
- shuttle freight trains, transporting trucks of a maximum weight of 44 tons.

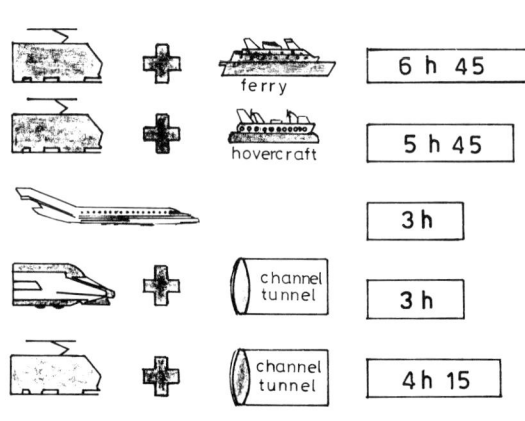

Fig. 1.19. Travel times between London-Paris with various transport modes

Fig. 1.19 illustrates the comparative travel times between London and Paris among the various transport modes: railway + ferry, railway + hovercraft, airplane, high speed + channel tunnel, ordinary railway + channel tunnel.

In 1990 it has been forecasted that for the year 2003 traffic through the Tunnel will reach 44.6 million passengers and 27.8 million of tons freight. Forecasts for the year 2013 are 53.9 million passengers and 38.0 million tons of freight.

More technical details about the Channel Tunnel Project are given in the relevant chapters (for instance soil mechanics of the project in chapter 3, section 3.2.2., etc.).

1.7. Aerotrain and Maglev

1.7.1. The Aerotrain

The aerotrain and magnetic levitation train technologies are based on guided vehicle transport (like conventional trains), but avoid any contact of the moving vehicle with the bearing structure on which transport is taking place, whereas railways rely on the metal (wheel) to metal (track) contact.

The aerotrain (see Photo 1.3, p. 3) is a vehicle running on a concrete bearing substructure in the shape of an inverted "T" (Fig. 1.20).

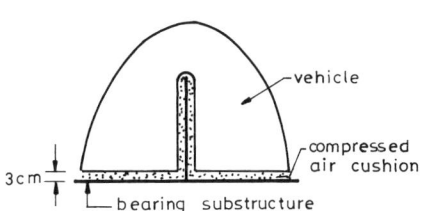

Fig. 1.20. The aerotrain principle

Propulsion is achieved without any wheel system, by a compressed air cushion blown between the vehicle and the bearing substructure. The aerotrain thus replaces the adhesion forces, necessary to propel conventional trains, by compressed air layers, (23).

The technology in question was developed in the 1960s in France and in 1969 had achieved the impressive speed of 422 km/h.

Even though there were various plans for aerotrain construction (e.g. Paris-Orleans where 18 km of bearing substructure had even been built, Brussels-Luxembourg, etc.), they were abandoned in the 1970s for various reasons, the main ones being:

- the new technique is not compatible with conventional railways
- construction proved much more expensive than a new conventional line (without the additional cost being offset by the much lower maintenance cost of the aerotrain compared with conventional railways)
- energy consumption (due to the air turbine used for aerotrain propulsion) is much higher than for conventional trains
- the carrying capacity of the aerotrain was low ($64 \div 96$ passengers in the prototype, but up to 160 passengers in two-car trains planned for later).

Secondary reasons, such as passenger safety considerations (possible fire in the vehicle which rides 5 m above ground), noise, questionable overall aesthetics, contributed to the abandonment of the plan.

1.7.2. Magnetic levitation trains (maglevs)

In magnetic levitation trains (See Photo 1.4, p. 3), contact between the bearing substructure and the vehicle is avoided, propulsion being ensured by magnetic phenomena. The fundamental principle of this technique is shown in Fig. 1.21.

Railway Engineering

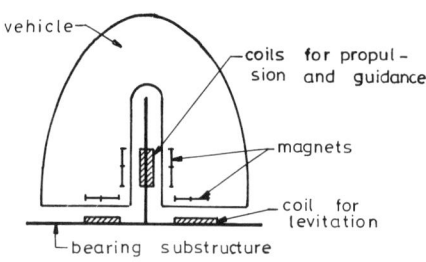

Fig. 1.21. The magnetic levitation principle

The bearing substructure is a concrete slab in the shape of an inverted "T" (or of a "U"). Suitably located magnets and coils generate the forces required for levitation, propulsion, and guidance. However, recent research has shown that the construction of a superconducting magnet fulfiling the above three requirements is possible, (39), (53). This technology was developed in the 1970s in Germany and Japan, where in 1979 during the course of testing a speed of 517 km/h has been attained. In Germany (Emsland), a magnetic-levitation test track was built with the following characteristics:

- bearing substructure of composite construction and superelevated section at a height of 5 m
- design speed 400 km/h
- cars 54 m long, weighing 120 tons, with a carrying capacity of 200 passengers
- minimum horizontal radius of curvature 4,000 m and maximum longitudinal gradient 10%.

The German Government has plans to construct a maglev line between Berlin and Hamburg. According to the plans, the maglev system will link the two cities (at a distance of 283 km) in less than one hour the year 2004, (53).

A magnetic-levitation test track of 42.8 km long is also under construction in Japan, with prospects of full maglev operation in 1990, (5). However, express links of airports with urban areas may be in the future candidates for the maglev technology.

1.8. Other transportation services with good prospects for the railways

High speeds is one area where the railways have comparative advantages to other transport means. Other such areas include urban rail services, combined transport, as well as transportation of bulk loads and, finally, integrated services which, in addition to transportation, involve the collection, storage, and delivery of goods (logistics).

1.8.1. Urban rail services

In an age with exploding traffic problems, the railways can decisively contribute to their alleviation by their large carrying capacities (Table 1.9). Many neglected railway lines connecting city centres to the suburbs are accordingly being modernized and used for urban rail services, thus relieving the traffic problem of the cities.

Railways and Transport

Table 1.9.
Carrying capacity of various transport systems, (47)

Transport system	Carrying capacity (in thousands of passengers per direction and hour)
Buses	0 — 4 ---- 6
Tramways	3 — 10 ----- 15
Light metro	8 ———— 20
Metro	10 ——————— 45
Suburban railway	10 ————————— 60

1.8.2. Combined transport

The various transport modes show comparative advantages as regards transport costs as a function of distance (Fig. 1.22). Thus, for short distances, the use of trucks is indicated, for intermediate distances railways have an edge, while great distances favour the use of ships. Increasing competition in the area of freight transport, however, makes the search for lowest cost compulsory. Several countries with important truck transit traffic (Austria and Switzerland, among others) set strict limits to the number of trucks in transit, so as to reduce congestion and saturation on the road networks. Finally, political events (such as the recent segregation of the former Yugoslavia) mandate the search for alternative, reliable and safe transportation routes. All the above have contributed to the growth of combined transport.

Combined transport may be defined as a composite transportation process involving at least two consecutive transport modes (e.g. truck-ship, train-ship, truck-train). Two main techniques were developed for combined transport:

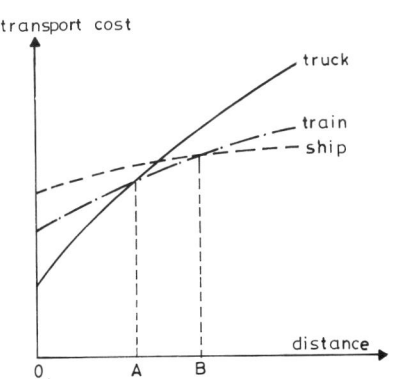

Fig. 1.22. Transport costs as a function of distance, for various transport modes, (41)

Photo 1.5. Containers

Photo 1.6. Ro-Ro combined transport

- Containers (Photo 1.5), used in road, rail, and sea transport. The tendency is to use containers as large as is allowed by the existing loading gauge, (48). Common container dimensions are 13.7 m long by 2.60 m wide.
- The Ro-Ro (Roll On - Roll Off, Photo 1.6) technique, whereby whole trucks or truck bodies with freight, are loaded on a train or ship, so that only a small portion of the trip is covered by road. According to EU regulations, the maximum dimensions of freight vehicles are: height 4.0 m, width 2.5 m, weight 40 tons.

Since combined transport requires vehicle transfer from one carrier to another (with associated expense), it is necessary to determine the minimum distance beyond which combined transport becomes cost-effective. The answer to this question is not simple, since it depends on the cost of labour, energy, the mechanical equipment for vehicle transfer, etc. Therefore, European conditions place this minimum distance at $700 \div 900$ km, whereas in the US it is set at 1,500 km, (49).

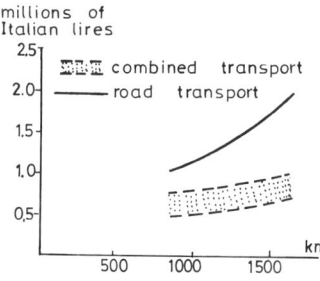

Fig. 1.23. Comparative cost of road and combined transport on the basis of Italian data, (49)

According to experience acquired in Italy, the comparative costs of road and combined transport are as illustrated in Fig. 1.23.

The development of combined transport necessitates the existence of a satisfactory road and rail network.

1.8.3. Bulk loads

In addition to combined transport, railways can further develop bulk load transport (raw materials, coal, petroleum, grain and other agricultural products, etc.). Railway competitivity in bulk load transport depends, among other matters, upon the marshalling yards facilities where freight trains are disassembled and reassembled and where long and (often) unjustified waits are occurring.

1.8.4. Integrated rail transport and logistics

Freight transport by rail has been until recently limited to carrying goods. The dynamics of modern transport, however, have broadened the scope of the transportation process. Reliable and speedy carriage is no longer sufficient. It must be also accomplished at the lowest cost, ensuring that a certain quantity of goods be made available at the required place and time. An important contribution to this effect has been recently achieved with freight transport system logistics, which involves the whole process encompassing timely information on the need to make available a particular item at a particular place and time, reliable and speedy transport, possible storage, and final delivery to the recipient, (49a). It is therefore clear that in this sense the transportation process has a much broader meaning. Besides, simple freight services can no longer meet the actual requirements of the production process, which in summary involve the following, (49), (49a):

- Demand and production do not occur at the same place, hence the need to move the particular product.
- Demand and production do not occur at the same rate. Whereas demand is generally characterized by regularity, although often seasonal, production, is carried out in large lots in order to minimize costs. The need therefore arises to store the product.
- Between the consecutive stages and before delivery of the product to its final destination, continuous and systematic logistics is necessary.
- Suitable adaptation of production to demand and optimum coordination of the consecutive stages between production and demand necessitate an integrated information forwarding system.

The main problem of the rail carrier therefore centres on the non-coinciding of the time and place of offer and demand for a particular product. Accordingly, a specific procedure making available the particular product at the time and

Railway Engineering

place of demand at the lowest overall cost should be sought. This can be solved through logistics charged with the organization, administration, and control of all processes between the production and delivery of a product, (49a).

From a logistics viewpoint, transport by rail show the following weaknesses which must be remedied so as to make it more competitive on the transport market:

- A fundamental weakness of transport by rail is 'idle times', while waiting to load the vehicle, as well as the fact that after a transport run, vehicles often return empty. Idle time may lead to unpredictable and long shipping rail times.
- When demand for transportation services is strong, rolling stock availability often proves inadequate. On the contrary, during idle periods, there is an excess of unused rolling stock.
- There is often a shortage of adequate storage facilities not meeting modern standards.
- Information is not transmitted fast enough between the requirement for transport, price quotation, actual transport and product delivery.

Most of the above weaknesses can be considerably alleviated by adopting logistic techniques so as to achieve the following:

- Reduction of idle times and of the number of vehicles coming back empty is mainly a matter of proper planning and of adapting the availability of transport services to demand and not vice versa.
- Insufficiency of the rolling stock during peak hours can be counteracted by rescheduling certain of these runs to other times.
- Storage facilities are a necessity for modern railway enterprises.
- Finally, the inflexibility in transmitting information and quick decision-making can be reduced by reorganization amendments using conventional, up-to-date, data processing, telematics and management techniques.

Figure 1.24 illustrates the evolution from simple transportation to logistics in the case of transport by rail.

1.9. International railway institutions

International railway cooperation is realized within the framework of the following international institutions:

1.9.1. The *International Railways Union (UIC)*, which was established in 1922 and currently has 87 members in 63 countries on five continents. Of the UIC members, 35 are considered active (networks in the Europe - Middle East geographical region), 32 are considered associated (networks not in direct communication with European networks), while another 20 are enterprises

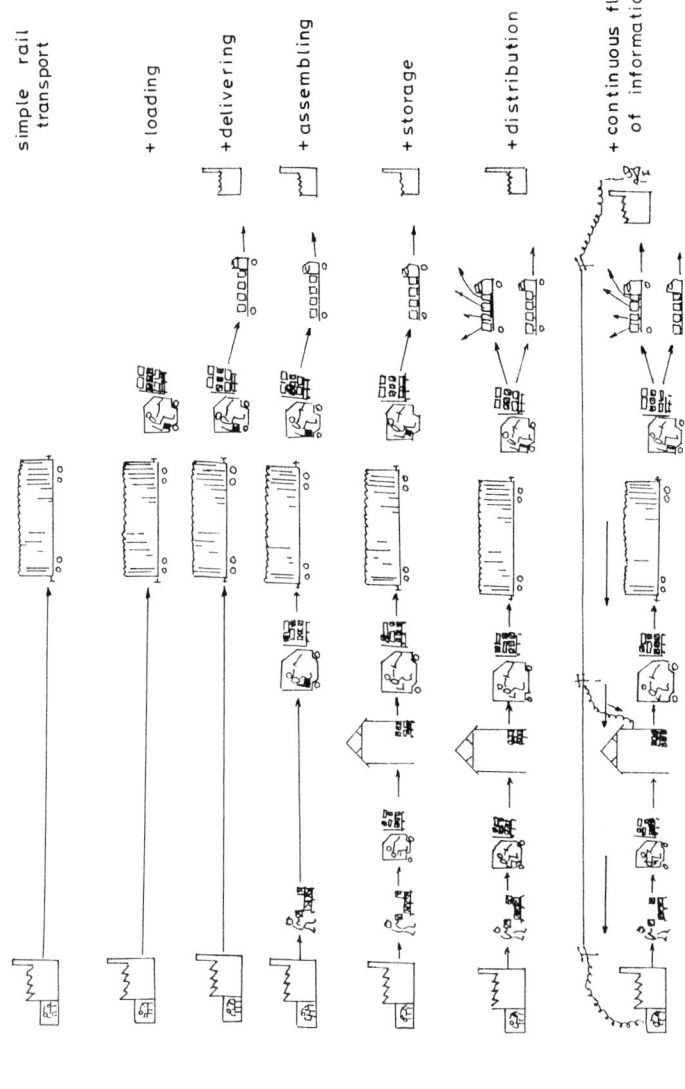

Fig. 1.24. From simple rail transport to logistics

connected by rail. The railway networks which are members of the UIC have a total track length of 550,000 km and a total workforce of 5,500,000 people.

The general objectives of the UIC are:

- Development of international railway transactions through the design and implementation of measures permitting railway services across national borders and ensuring quality in both passenger and freight traffic.
- Informing international organizations, decision centres, and public opinion on the usefulness and advantages of transport by rail.

Within this general framework, UIC activities cover the following sectors:

- allocating income and offsetting debit between networks
- planning the integration and rationalization of the technical equipment, exploitation methods, data processing, etc.
- research on new technological advances concerning track equipment, rolling stock, etc.
- statistics and other information.

1.9.2. The *European Conference of Ministers of Transport (ECMT)*, which is an agency working in close collaboration with the Organization for Economic Cooperation and Development (OECD).

1.9.3. The *Community of Railway Networks* of the European Union member-countries, a body operating within the framework of the UIC and the European Union. It aims at establishing common positions and policies of railway networks in European Union member-countries.

1.9.4. The Railway Transportation Committee of the *United Nations Economic Committee for Europe*, with the participation of delegates from the various governments.

1.9.5. The *Research Institute* (known with the initials *ORE* of its former French name) which is an agency of the International Railways Union aiming to organize and coordinate research and test procedures advancing railway technology. Topics investigated are divided into the following five categories (denoted by the letters A, B, C, D, E):

    A: Traction, signalling, telecommunications
    B: Rolling stock
    C: Interaction between rolling stock and track
    D: Track, bridges, tunnels
    E: Materials' technology.

Throughout the decade 1980-1990, activities of the Research Institute of UIC focused on the following basic areas many of which are continuing in the 1990s:

Railways and Transport

- high speeds
- increase of the load per axle from 20 to 22.5 t, and of the speed of freight trains to 120 km/h and for certain routes to 160 km/h
- standardizing rolling stock
- signalling studies
- studies on environmental impact (transmission of railway vibrations to the environment, etc.)
- new lighter and more efficient materials.

2 The Track System

2.1. Areas of railway engineering: track, traction, operation

As the analysis in chapter 1 shows, railway engineering aims at inter-disciplinary knowledge and requires the competences of the civil engineer, the electrical engineer, the mechanical engineer and the economist. Beginning with the railway network organization, it has become customary to distinguish railway engineering into three topic areas:

- *Track topics.* Subjects of railway bearing infrastructure are dealt with, in order to ensure the safe operation of the rolling stock at the forecasted speed. The superstructure (rails, sleepers, ballast) and the subgrade are central subjects of track topics. Track topics also include railway stations and level crossings.
- *Traction topics.* Subjects concerning rolling stock are elaborated on. Traction topics also include electric traction, telecommunications, and signalling. Certain networks, however, include these latter in the area of track topics, since they are part of the permanent railway infrastructure.
- *Operation topics*, which include:
    * Commercial operation, in which commercial and pricing policies are analyzed.
    * Technical operation, where issues concerning schedule organization, optimum use of rolling stock and traffic safety are examined.

To the above should be added the topics on metropolitan subways (metros), which constitute a specific railway class of their own of great importance to mass transit in large urban centres.

This book contains mainly civil engineering railway aspects and thus it is principally devoted to track topics. However, railway stations are not analyzed, since they belong to the architectural engineering domain. Given that electrification is a component of the permanent way and that train dynamics is necessary for the analysis of various aspects of the track and the rolling stock, they are included in this book (chapters 12 and 13).

The Track System

2.2. The components of the track-subgrade system

Two discrete subsystems are distinct in a railway line (Fig. 2.1), (55):
- The superstructure or track (rails, sleepers, track support) which supports and distributes train loads and is subject to periodical maintenance and replacement.
- The subgrade (base, formation layer), on which the train loads, after adequate distribution, are transferred and which in principle should not be subjected to intervention during periodical maintenance of the railway track.

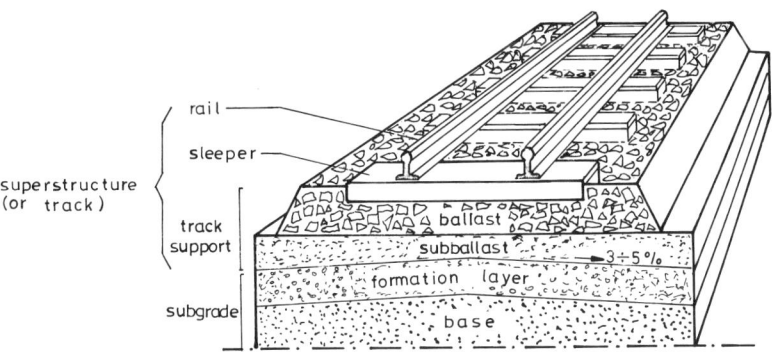

Fig. 2.1. The track-subgrade system

Photo 2.1. Railway track

The superstructure is composed of:
* The rails, which support and guide the train wheels.
* The sleepers (also called ties, principally in North America) with their fastenings, which distribute the loads applied to the rails and keep them at a constant spacing.
* The ballast, usually consisting of crushed stone and only in exceptional cases of gravel. The ballast should ensure the damping of most of the train vibrations, adequate load distribution and fast drainage of rainwater.

Railway Engineering

* The subballast, consisting of gravel-sand. The subballast protects the subgrade top from the penetration of ballast stones, while at the same time further distributing external loads and ensuring the quick drainage of rainwater.

In the subgrade the following are distinguished:
* The base, which in the case of the track laid along a trench consists of on-site soil, while in the case of an embankment is composed of soil transported to the site.
* The formation layer of the base, used whenever the base soil material is not of adequate quality.

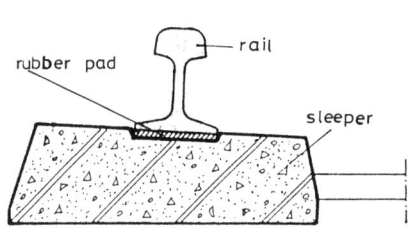

Fig. 2.2. Rubber pad between rail and sleeper

The depth to which disturbances from train passage are felt extends to around 2 m below the subgrade top, (55), and this is the depth down to which will henceforth be referred to by the term subgrade.

Resilient pads are placed between rail and sleeper to further attenuate train vibrations (Fig. 2.2), (60). Two pad thicknesses are usually applied: 9 mm and 4.5 mm.

The succession of the various layers of the track system is characterized by a gradual increase in the surface area as we proceed to lower layers and by a considerable reduction of the developed stresses, (4). Accordingly, stresses are reduced by 1,000 to 5,000 times, between the point where the wheel load is applied and the subgrade, (Fig. 2.3).

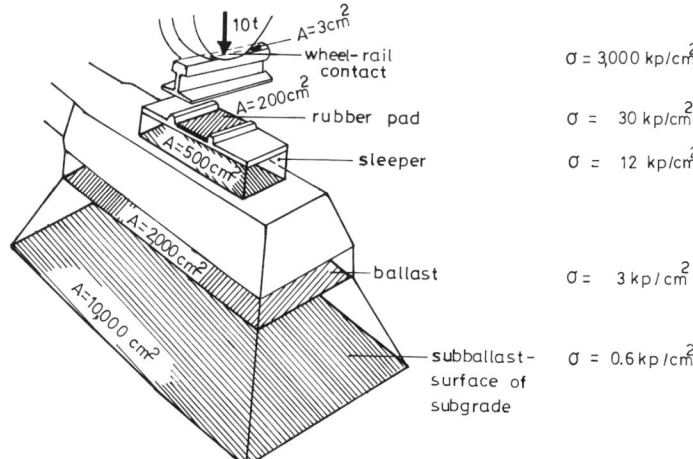

Fig. 2.3. The base area of each component of the track system and the distribution of train load, (4)

The Track System

2.3. Track on ballast and concrete slab

The track usually rests on ballast, in which case we have a flexible support or a ballasted track, (Fig. 2.4.). However, it is possible, that the track is supported by a concrete slab, in which case there is an inflexible support or slab track. Although an inflexible support is often used in certain rail networks (e.g. the Japanese and the German networks), it is most effective when used in tunnels, because it allows a smaller cross-section and facilitates maintenance. In most cases, a ballasted track is preferable, because it ensures flexibility (an important factor in the event of differential settlements) and much lower construction cost, while at the same time offering a very satisfactory transverse resistance, even at very high speeds, (54), (55), (143). The problem of noise, which is much higher with track on concrete slab than with track on ballast, should not be disregarded. When a slab track is applied (e.g. in the case of a tunnel), the sudden variation in track stiffness (felt by passengers as a jolt) is lessened by placing rubber pads of a suitable thickness along the tunnel entrance and exit. Slab track is examined in more detail in chapter 6, section 6.8.

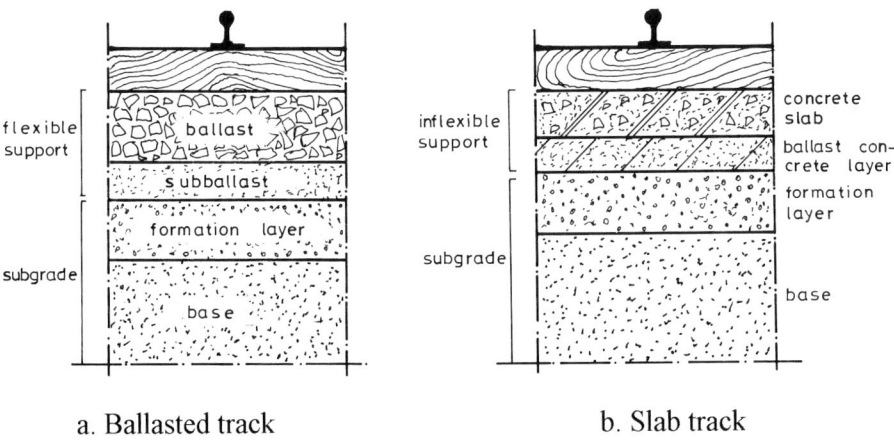

a. Ballasted track   b. Slab track

Fig. 2.4. Ballasted track and slab track

2.4. Track gauge

The track gauge is defined as the distance between the inner sides of the rails, measured 14 mm below the rolling surface (Fig. 2.5). Tracks with different gauge values have been laid, as follows:

Railway Engineering

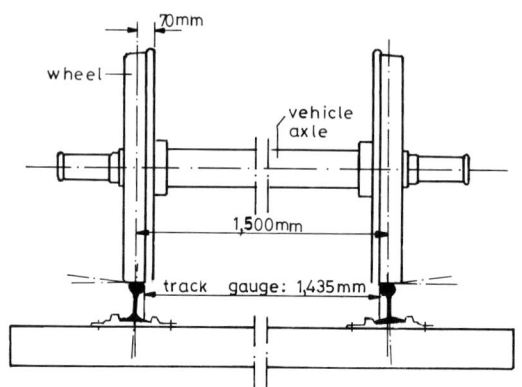

Fig. 2.5. Track gauge

- *Standard* gauge, e = 1.435 m. Most lines have been laid at this gauge, found to optimize rolling stock dimensions. In standard gauge lines, the maximum permissible deviations from the 1.435 m value range from +10 mm to -3 mm.
- *Metric* gauge, e = 1.000 m or e=1.067 m. Usually secondary lines are laid using the metric gauge.
- *Broad* gauge, e=1.524 m (Russia), e=1.672 m (Spain) and elsewhere. They have been constructed so as to differentiate them from the standard gauge, mainly for political reasons, to prevent standard-gauge rail vehicles from trespassing into these networks.

It should be noted that gauge values had initially been expressed in British measurement units (inches), hence the general irregularity of the above numerical values by their conversion into metric units.

On curves below 400 m radius it may be necessary to increase the track gauge to ease movement of vehicles and reduce wheel and rail wear (see chapter 11, section 11.3.4, table 11.1).

2.5. Load per axle and traffic load

2.5.1. Load per axle

The load per axle and the traffic load (tonnage) running on the line are critical factors for track and subgrade fatigue. Depending on track equipment, different values of axle load may be applied on the various lines, classified into four categories:

    A : Maximum load per axle 16 t
    B : Maximum load per axle 18 t
    C : Maximum load per axle 20 t
    D : Maximum load per axle 22.5 t.

Category D was derived by increasing the load per axle of category C from 20 to 22.5 tons, in an effort to reduce the operating cost, especially for freight transport. This increase was made after years of research and studies, (63), with controversy which did not focus as much on track strength as on the behaviour of bridges which had been designed for a 20 t load per axle on the basis of simplified theories of elastic behaviour. Research on the elastoplastic behaviour of materials, (63), has shown that bridges designed for loads of 20 t per axle can withstand loads of 22.5 t per axle without the need of strengthening, due to reserves which the elastic theory had failed to take into account.

Certain networks, however, use larger per-axle loads. In the U.S. (where railways are mainly limited to freight transport) the maximum load per axle is 25 ÷ 32 t, while in Russia (where broad-gauge lines are used) it is 25 t.

A series of studies has shown, (56), that rail fatigue is an exponential function of the per-axle load Q, and stresses developed within the rail are proportional to the parameter Q^a, where the exponent a takes values in the range of 3 to 4 and closer to 4. Thus, any increase in the load per axle results in a much larger increase in track material fatigue.

2.5.2. Traffic load

Various kinds of rail vehicles are running on a railway track: passenger vehicles, freight vehicles, main-line locomotives, shunting engines, etc. The algebraic sum of the vehicle loads cannot give an accurate picture of the running load, because it does not take into account the way in which the load is applied, the running speed, etc. Therefore, a complex parameter giving an accurate estimate of the passing traffic load is necessary. Railway engineering uses the analogue of the Passenger Vehicle Unit (PVU) of traffic engineering. In order to determine the traffic load (or tonnage) on a track, the loads of the various trains are converted into the equivalent passenger train loads.

The theoretical load T_{th} of the track is firstly calculated from the equation

$$T_{th} = T_p + k_{fr} \cdot T_{fr} + k_{tr} \cdot T_{tr} \qquad (2.1)$$

where T_p : daily passenger vehicle traffic
T_{fr} : daily freight vehicle traffic
T_{tr} : daily traction engine traffic
k_{fr} : 1.15
k_{tr} : 1.40

The traffic load T of the line is thereafter calculated, taking into account the train running speed, from the equation:

$$T = S \cdot T_{th} \qquad (2.2)$$

where S = 1.0 for lines without passenger traffic
S = 1.1 for lines with mixed traffic and $V_{max} < 120$ km/h
S = 1.2 for lines with mixed traffic and $120 < V_{max} < 140$ km/h
S = 1.25 for lines with mixed traffic and $V_{max} > 140$ km/h.

Railway Engineering

Based on the daily traffic load, the various railway lines are classified into groups according to UIC standards (Fig. 2.6).

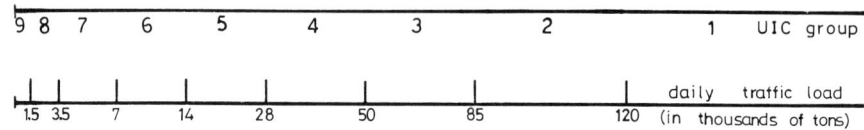

Fig. 2.6. Classification of railway lines into UIC groups according to their daily traffic load

2.6. Sleeper spacing

The study of track behaviour has shown that, the closer the sleepers are spaced, the better is the load distribution and the smaller are the stresses developed. As sleeper spacing is made smaller, however, track maintenance becomes more difficult. A compromise should therefore be found between the above two requirements.

Sleeper spacing is defined as the distance between the axes of consecutive sleepers, and its optimum value for standard gauge lines is 0.60 m, which can be reduced to 0.55 m in cases of subgrade instability and small radii of curvature. Occasionally the number of sleepers per km is used as a parameter, with 1,666 sleepers/km as the average value. In networks with higher values of load per axle (USA, Russia), sleeper spacing may be reduced to 0.50 m. On light-weight railways sleeper spacings may be increased, but rail fatigue must be carefully considered.

2.7. The wheel-rail contact

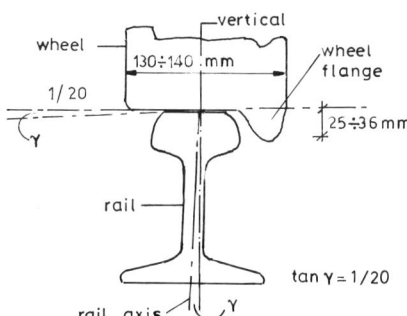

Fig. 2.7. The wheel-rail contact

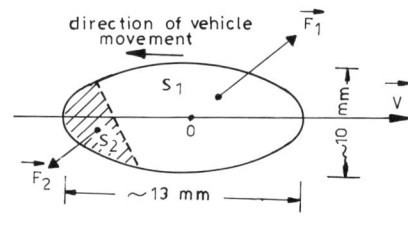

Fig. 2.8. Detail of the wheel-rail contact surface

The Track System

A fundamental characteristic of rail vehicles is wheel movement guided by the two rails. Wheel-rail contact (Fig. 2.7) has an elliptical form (Fig. 2.8), (59). The rail axis inclination to the vertical is termed conical tread. Conical tread usually has the value 1/20. However, after the general adoption of rail type UIC 54 and 60 (see chapter 5, section 5.2), conical tread in certain networks was reduced to a value of 1/40, (61).

Wheel movement on the rail gives rise to the creep effect. Indeed, the wheel-rail contact surface can be divided into two areas, S_1 and S_2, the sizes of which depend on the vehicle speed and with different effects taking place in each area. Thus, the vehicle rolling resistance consists of two components, F_1 and F_2, corresponding to areas S_1 and S_2 respectively and of opposite direction. F_1 is generated by vehicle movement, i.e. it is of kinematic origin, while F_2 is generated by elastic deformation of the S_2 surface, i.e. it is of elastic origin.

As speed increases, S_1 becomes larger and S_2 respectively smaller. At *high speeds*, S_2 almost decreases to zero, and therefore the rolling resistance of the vehicle coincides with dynamic friction. According to Coulomb's law, the following relation will then apply:

$$F = \Phi = \mu Q \qquad (2.3)$$

where F : vehicle propelling force
 Φ : vehicle friction
 μ : coefficient of friction
 Q : vertical load of wheel

At *low speeds*, on the contrary, we have creep effects and Coulomb's law no longer holds. In this case it is possible to make the simplifying assumption that the propelling force is proportional to the ratio of the sliding speed to forward speed.

$$u = \frac{\text{sliding speed}}{\text{forward speed}}$$

If f is the proportion factor, then at low speeds

$$F = f u \qquad (2.4)$$

The following equation gives a better approximation, than the above:

$$\frac{1}{F^n} = \frac{1}{fu^n} + \frac{1}{\Phi^n}, \qquad n \neq 2 \qquad (2.5)$$

Railway Engineering

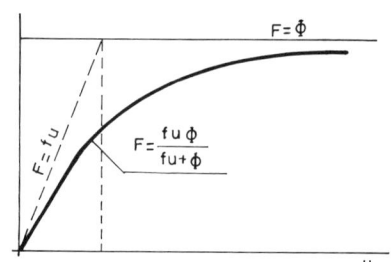

Fig. 2.9. Propelling force at intermediate speeds

At *intermediate speeds*, experimental data on the propelling force F have yielded the following relation

$$F = \frac{fu\Phi}{fu + \Phi} \quad (2.6)$$

Lines $F = fu$ and $F = \Phi$ are tangent to the curve of equation (2.6).

A closer approximation of creep forces at the wheel-rail contact surface has been given by Kalker, (61a).

According to the theory developed by him, the elliptical contact surface may be divided into two sections:
- The first section of the contact surface undergoes creeping and each point of the first section transmits to the second part of the contact surface a transverse force given by Coulomb's relation.
- The second section of the contact surface leads to adhesion (at zero creep values) and the forces transmitted by the second section to the first have a lower value than that given by the Coulomb equation.

Conventional railways use metal wheels. Rubber wheels started being used after 1970 in subways to reduce vibration transmitted to the environment and increase acceleration and deceleration. Rubber wheels do not permit high speeds and are subject to deterioration under bad weather conditions. For this reason they are used only in subway vehicles.

2.8. Transverse wheel oscillations along the rail

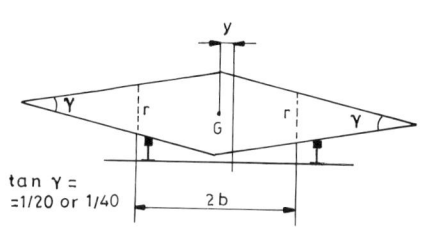

A rail vehicle can be simulated by a solid composed of two cones connected at their base (Fig. 2.10). This solid is supported by the two rails and the angle γ of the cones is equal to the wheel conical tread, $\tan\gamma = 1/20$ or $1/40$.

Due to the conical tread, the wheel follows a snaking path along the rail (Fig. 2.11).

Fig. 2.10. Simulation of a rail vehicle by a solid composed of two cones

The Track System

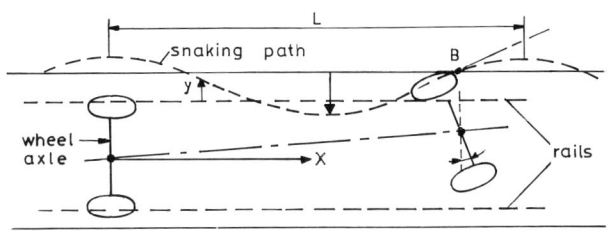

Fig. 2.11. The snaking path of the wheels along the track

The gap between the rail head and the wheel allows the latter to move transversely, which causes the snaking movement of the rail vehicle. Transverse wheel movements are opposed by creep forces.

Transverse movements were analyzed by Klingel, (4), and this phenomenon is often referred to by his name. Klingel presented a kinematic analysis of this phenomenon, assuming a sinusoidal transverse movement with no attenuation. Let (Fig. 2.12):

- y : the transverse movement from equilibrium position
- v : the travelling speed
- s : the track gauge
- γ : the wheel conical tread
- R : the radius of curvature of the snaking path
- r : the wheel radius at the equilibrium position
- x : the abscissa

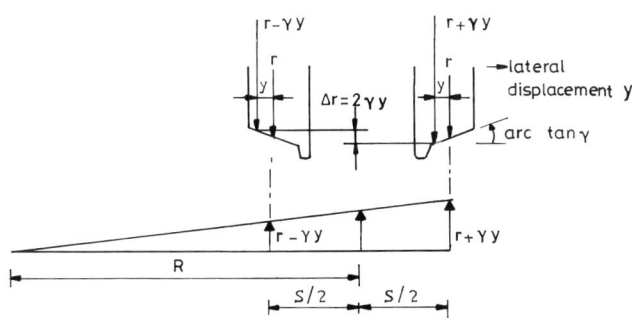

Fig. 2.12. Analysis of snaking wheel movements according to Klingel

From Fig. 2.12 and the similar triangle relationship, it follows that:

$$\frac{r + \gamma y}{r - \gamma y} = \frac{R + s/2}{R - s/2} \qquad (2.7)$$

From kinematics it is given that:

$$\frac{1}{R} = -\frac{d^2y}{dx^2} \quad (2.8)$$

From equations (2.7) and (2.8) we derive the differential equation for the snaking movement:

$$\frac{d^2y}{dx^2} + \frac{2\gamma}{rs} y = 0 \quad (2.9)$$

Given the limit condition

$$y(0) = 0 \quad (2.10)$$

the solution for the differential equation is:

$$y = y_o \sin 2\pi \frac{x}{L} \quad (2.11)$$

with y_o the amplitude and L the wavelength of the snaking movement

$$L = 2\pi \sqrt{\frac{rs}{2\gamma}} \quad (2.12)$$

The maximum value of the transverse acceleration is:

$$\gamma_{max} = \frac{d^2y_{max}}{dx^2} = 4\pi^2 y_o \frac{v^2}{L^2} \quad (2.13)$$

As a numerical example, let $r = 0.45$ m, $s = 1.435$ m, $\gamma = 1/20$, in which case $L = 16$ m. If, however, $\gamma = 1/40$, then $L = 22$ m.

The frequency of the snaking movement can be found from the relation

$$f = \frac{v}{L}$$

When frequency f is the same as the frequency at which the rolling stock resonates, then wheel movement becomes instable. The transverse acceleration, which is a measure of the forces exerted, shows the opposing effects generated by increasing speed and decreasing the transverse movement wavelength. A conical tread of 1/40 instead of 1/20 is therefore better at the same speed. Conversely, as the wheels gradually wear off, conical tread increases and as a result wavelength decreases.

However, in modern rail vehicles the rolling stock body is not directly supported by the wheel axles but by bogies, which are in turn supported by the

2.9. Rail mounting angle on sleeper

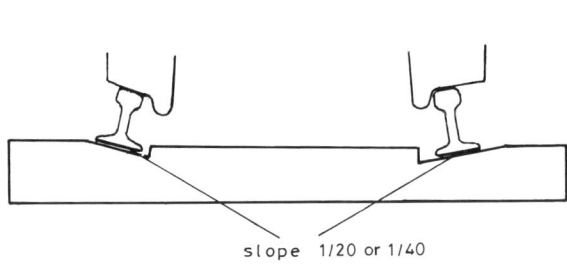

Fig. 2.13. Rail mounting slope on sleeper

Due to the conical tread, rails are mounted on sleepers at a slope. As explained above, the conical tread is usually given the value 1/20. A reduction of the value of the conical tread has been suggested, however, especially at high speeds. Several networks are already mounting the rails on the sleepers at a slope of 1/40, (61).

2.10. Load gauge

The *load gauge* is defined as the minimum external border required to remain free around the rolling stock. The load gauge is distinguished in:
* *static* load gauge, which is the minimum external border required to remain free as long as the train is not moving,
* *dynamic* load gauge, which is the minimum external border required to remain free while the train is moving. The boundary enclosing the clear spaces required around the dynamic load gauge is the *structure* gauge.

The load gauge mainly depends on two parameters:
* the rolling stock width (usually between $2.60 \div 3.30$ m)
* the spacing between the axes of the two tracks.

The International Railways Union (UIC) has specified the load gauge which is required to ensure that trains from one network can run on other network tracks without any problems (Fig. 2.14). The distance b between the axes of the two tracks may vary between 3.57 m and 3.67 m according to the UIC, depending on the maximum acceptable running speed. Even with the UIC standardization, however, significant differences in load gauge are observed, mainly in the U.K. (Fig. 2.15), where load gauge has smaller dimensions than in Continental Europe. American load gauge (Fig. 2.16) has also significant dimension differences compared to the European ones.

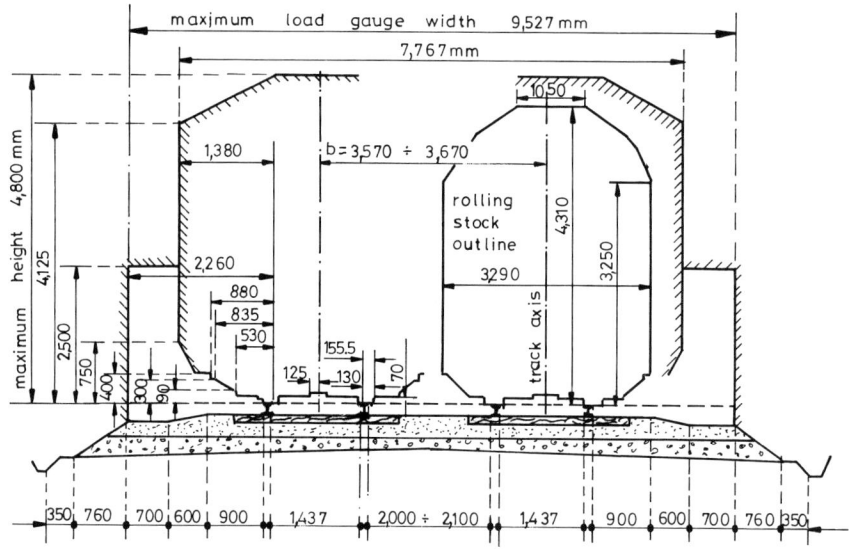

Fig. 2.14. Medium- and low-speed train load gauge (according to Regulation 505 of the UIC)

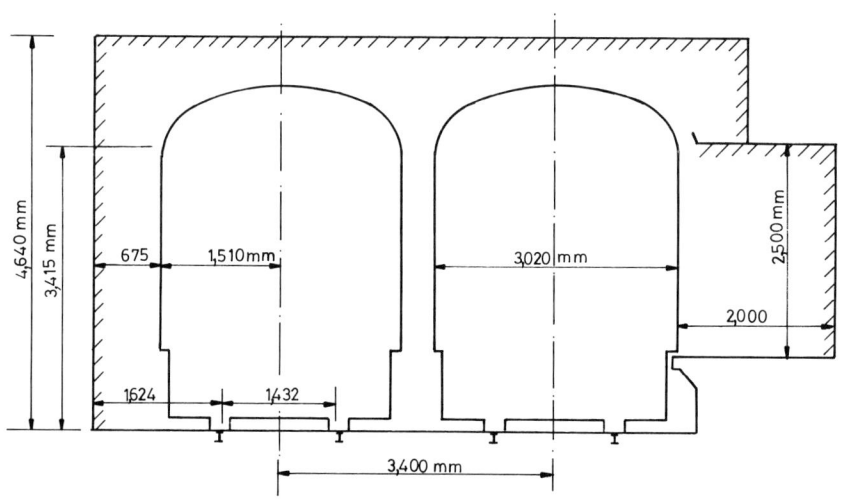

Fig. 2.15. British load gauge

The Track System

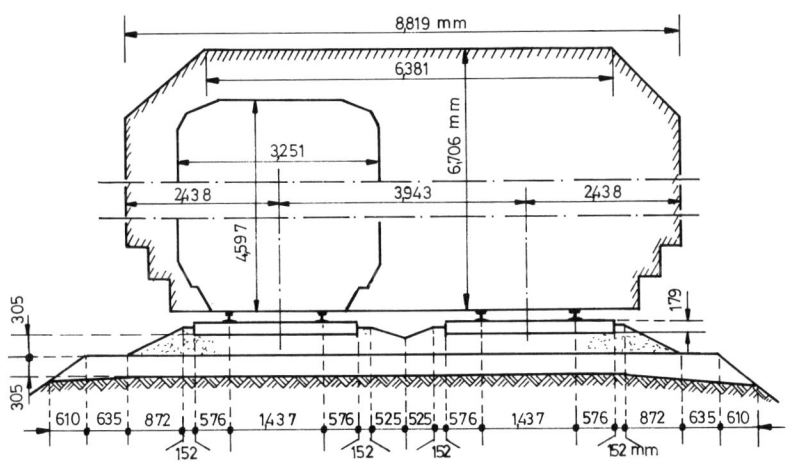

Fig. 2.16. American load gauge

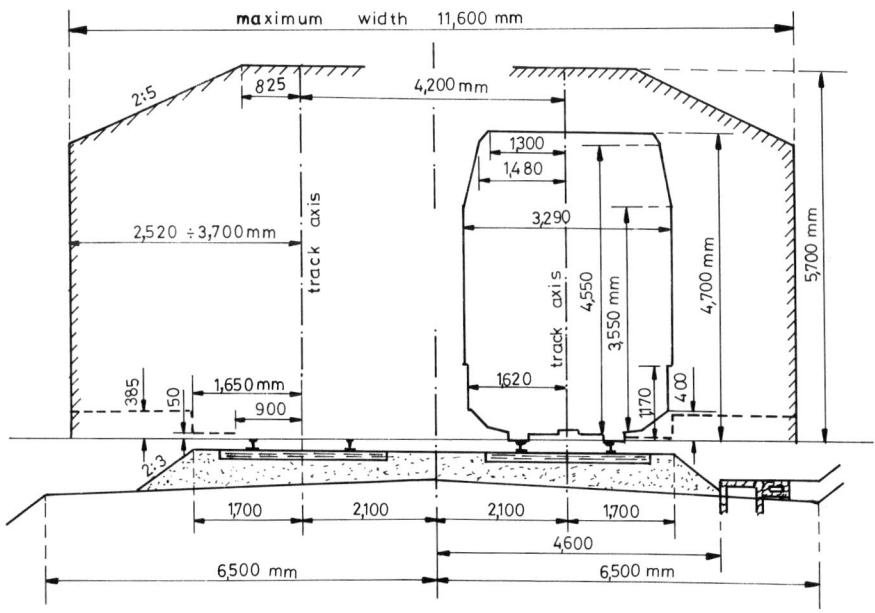

Fig. 2.17. High-speed train load gauge (French TGV)

Railway Engineering

The load gauge is different for high-speed trains, mainly because of the large spacing necessary between the axes of two tracks, as well as the large lateral spacing (Fig. 2.17).

The dynamic load gauge requires special attention when trains are running through tunnels, as well as in the case of metropolitan subways (Fig. 2.18). Each railway and metro Authority must have its own local structure gauge requirements which must be followed in each specific case.

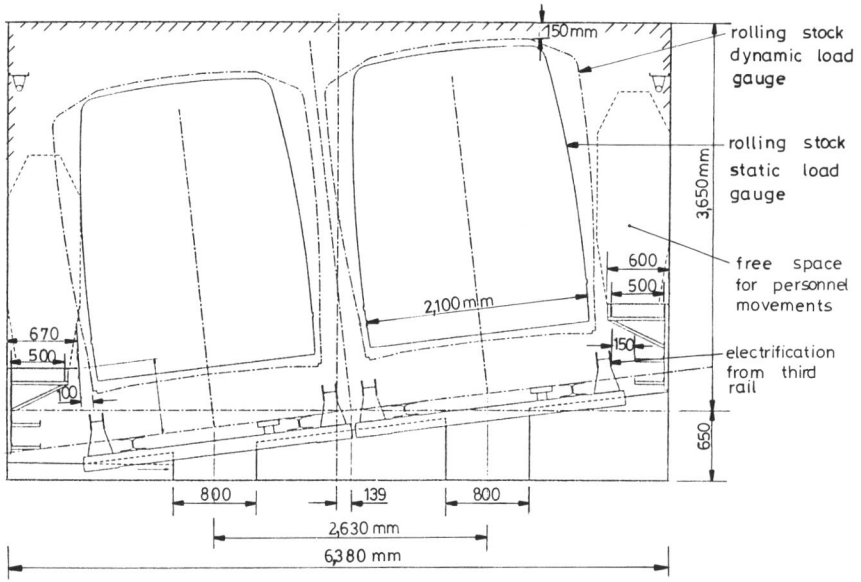

Fig. 2.18. Dynamic and static load gauge of a metro on curved track

2.11. Forces generated by rail vehicle movement - static and dynamic analysis

2.11.1. Forces generated

Forces exerted on the track during the running of a rail vehicle may be classified depending on their direction in:
- *Vertical* forces. These are the cause of mechanical stresses in the track. When subjected to vertical forces, the behaviour of certain parts of the track (rails, sleepers) is elastic, while that of the ballast and the subgrade is elasto-plastic, (55), (84). Vertical forces are critical to the dimensioning of the various components of the track system.

- *Transverse* forces. These influence train running safety, and may, under certain conditions, cause train derailment. Transverse forces effects are analyzed in chapter 8.
- *Longitudinal* forces. These are generated by acceleration and deceleration during train operation. Longitudinal forces are taken into account in the design of bridges on railway lines.

Although an accurate analysis of the various phenomena has shown a non-linear behaviour, the inaccuracy introduced by the omission of the non-linearity is often smaller than the inaccuracy introduced by other calculation parameters, e.g. the values of the mechanical characteristics, (55), (56). The method prevailing in railway engineering involves separately analyzing vertical, transverse, and longitudinal phenomena generated during train operation, which implies that the effects are assumed to be linear. It is an approximation which the engineer must be aware of in the analysis of the various effects.

2.11.2. Static and dynamic analysis

A frequent assumption in railway engineering is that both the wheel and the rail are free of defects. Measurements of the stress quantities have furthermore shown that the influence of time may be considered negligible. In such conditions, a static analysis of the various effects is adequate.

In both the wheel and the rail, however, defects do occur, causing additional dynamic loads to the wheel-rail system. These additional dynamic loads become more important as train speed increases. Force measurements have shown that, at loads of 10 t per wheel and 200 km/h speeds, the additional dynamic load is equivalent to increasing the static load per wheel by 6 tons, (56). Therefore, if at low speeds the additional dynamic loads can be neglected, this is not so at medium speeds and even less so at high speeds (see also chapter 4, sections 4.4 and 4.5).

Due to their random nature, an accurate analysis of the additional dynamic loads is possible by spectral analysis, (57), (62). With this method it was found that additional dynamic loads may be classified into two groups:
- Additional dynamic loads caused by *sprung masses* (*rolling stock*) and influenced by the type and characteristics of the rolling stock. Sprung mass oscillations increase with train speed, but at a lower rate. The increase of the oscillations of the sprung masses is a function of their vertical oscillation resonance frequency. The influence of this resonance frequency is considerable.
- Additional dynamic loads caused by *unsprung masses* (*wheels, rails*), which are proportional to speed, the magnitude of track defects, the square root of the unsprung mass, and the square root of the vertical stiffness of the track. The standard deviation, sd, of the additional dynamic loads ΔQ caused by the unsprung masses may be expressed by the relation, (56):

Railway Engineering

$$sd_{\Delta Q} = V\sqrt{\frac{A\,m\,h}{2\alpha}} \qquad (2.14)$$

where V : vehicle speed
- m : unsprung mass per wheel
- h : vertical stiffness of the track, which as explained in chapter 4, section 4.1.2, is defined as $h = \frac{Q}{z}$, with Q the load per wheel and z the vertical settlement at the rail level
- α : damping factor
- A : empirical coefficient depending on track maintenance conditions.

2.12. Influence of the forces generated on passenger comfort

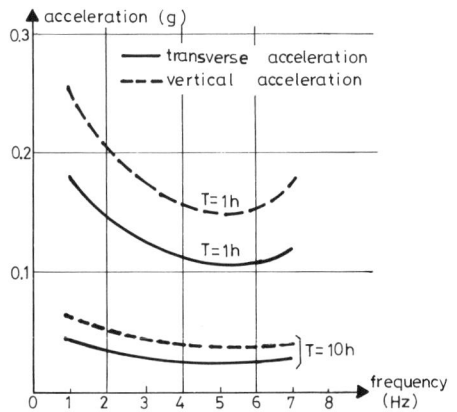

Fig. 2.19. **Sensitivity of the human body to vibration and curves of equal physical comfort**

Passenger comfort is affected to a large extent by the exerted forces which determine the value of the acceleration to which the human body is subjected. The sense of comfort, however, is also affected by the vibration frequency. It was found, (1), that comfort is minimum at frequencies in the order of 5 Hz and that the human body better accepts vibrations corresponding to frequencies higher than 5 Hz and up to 20 Hz, (Fig. 2.19).

2.13. Construction cost of a new railway line

The construction cost of a new railway line is influenced by several factors:
- Layout characteristics, mainly the number and size of bridges and tunnels. It should be noted that in railway lines of comparable level, the existence of many civil engineering structures (tunnels, bridges) may double or even triple the construction cost.
- The expropriation cost which, especially in urban areas, may considerably inflate the construction cost.
- Labour costs, which vary from country to country (often within the same country).

The use of cost data based on information from other countries' analyses should therefore only serve as a rough estimate of the various cost parameters, always keeping proportions in mind. Thus, the construction cost per km of the South-East TGV Paris-Lyons line (including relatively little civil engineering) was FF 21 million in 1985 prices, while the corresponding cost of the Atlantic TGV Paris-Bordeaux line (with comparatively more civil engineering) was FF 31 million in 1985 prices. The cost per km of the new IC Hannover-Würzburg line with many and difficult civil engineering projects, amounted to, in 1983 prices, DM 35 million, (44).

Finally, the construction cost distribution of a new railway line to the various components of the railway system differs greatly with each case. Table 2.1 illustrates the average values from French, Spanish, Greek and German data for lines with no major civil engineering projects.

Table 2.1.
Typical distribution values of construction cost of a new railway line to the various components of the railway system

Subgrade	45÷30%
Civil engineering projects	10÷25%
Track	20%
Signalling - Telecommunications	10%
Electric traction	10%
Design	4÷5%

3 Railway Subgrade

3.1. The importance of the railway subgrade on track quality and its functions

Railway subgrade is particularly important in ensuring that track quality reaches the standard necessary for the safe and comfortable running of trains. Railway networks make serious efforts to improve passenger comfort. These efforts, however, mainly concentrate on the railway track and often disregard the fact that many problems appearing at track level are traceable to the subgrade rather than to the track structure above.

It should be stressed that, in the past, studies concerning the subgrade were influenced by ideas prevailing in highway engineering. This had the advantage of using the technical experience acquired with highways, but the disadvantage, when highway design specifications were applied literally, that the techniques implemented were not compatible with the peculiarities of the railway environment.

The railway subgrade problem arises in different ways in new and existing track layouts. Accordingly, in new layouts the subgrade design is a function of the forecasted loads (load per axle and track tonnage), sleeper type and ballast thickness. A rational consideration of the problem requires that various parameters defining the subgrade be taken into account: soil type, hydrogeological conditions, mechanical strengths.

On the other hand, in existing layouts, the problem is different. The policy of the railway networks (higher speeds, higher loads per axle) leads to increased subgrade stresses. As in existing layouts, the lower surface of the subballast and the upper surface of the subgrade have formed a compact zone, which should be disturbed as little as possible, there is little possibility of intervention to the subgrade. Any intervention to the subgrade should be limited to areas where particular problems have arisen, and should be scheduled as much as possible to be performed during periodical track maintenance. The decision between subgrade improvement and an increase in the ballast layer thickness should be the subject of a technical and economic study, and therefore is difficult to make in advance, (64).

Therefore, the railway subgrade should fulfil the following functions:
- enable passenger and freight trains to run safely at design speeds
- support the heavy axle loads imposed by freight trains
- minimize future track maintenance costs.

These functions can be achieved by:
- limiting settlements of the original ground and consolidation within the embankment filling
- providing an arrangement that will be stable under the imposed railway loadings and the weight of the earthworks
- ensuring that the condition of the formation does not deteriorate during its working life.

3.2. Geotechnical analyses and soil classifications

3.2.1. Analytical geotechnical study

It would be ideal if a thorough geotechnical study of the ground were completed before the construction of the railway subgrade. Such a geotechnical survey is costly, therefore it is performed mostly before constructing new railway lines.

A geotechnical investigation should indicate:
- whether material for embankment construction is available on site or will require fill to be imported,
- where weak ground requires treatment before filling can commence,
- where ground water levels may cause problems,
- the measures necessary to ensure the stability of earthwork slopes in the long term,
- where cuttings require particular drainage or protective measures,
- the appropriate type of plant to be used for cut and fill operations.

Figure 3.1 illustrates the geotechnical characteristics along the Channel Tunnel, which is constructed along a layer of blue chalk that was proven to resist water penetration.

3.2.2. Geotechnical classifications of soils

However, in existing railway lines such an analytical geotechnical survey is not necessary. Nevertheless, a general knowledge of the basic parameters of the mechanical behaviour of the ground is essential. The various geotechnical classifications, adopted mainly for highway engineering projects, are a helpful indication in this purpose. These classifications are based on the following characteristics: granulometric grade and Atterberg limits (liquidity limit, plasticity limit, shrinkage limit). Occasionally, mechanical parameters are also taken into consideration, such as the CBR index, etc.

Railway Engineering

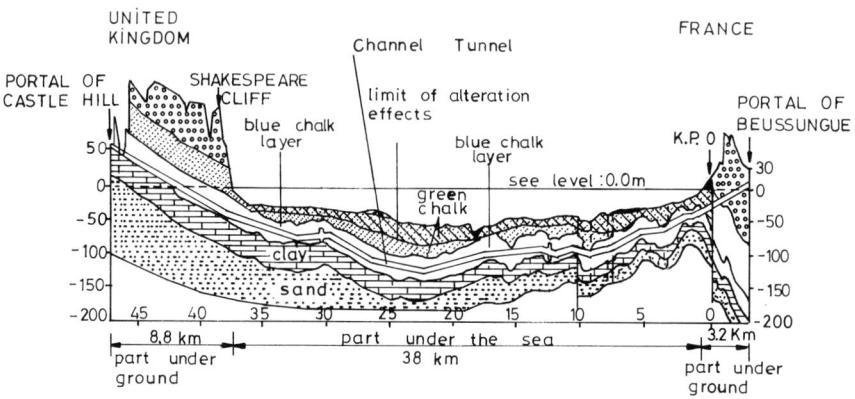

Fig. 3.1. Geotechnical characteristics along the Channel Tunnel

Various railway networks classify soils differently, as illustrated for the following countries, (64) (66):
- the UK, France, Germany, Switzerland and others use the USCS (United Soil Classification System) classification, also known as Casagrande classification,
- Scandinavian countries mainly rely on granulometric grading of the material,
- Italy, Greece and others use the AASHO (American Association of State Highway Officials) classification.

Soils composed of mixtures of two or more groups of fine-grain sizes are usually considered separately. An accurate classification of such soils with similar granulometric compositions requires that plasticity characteristics (Casagrande diagram) be also taken into consideration.

Despite the small differences due to different classification methods, the following terminology is commonly acceptable in soil engineering:
- Rock: low-, medium-, or high-variability rock, depending on the decay-disintegration it has undergone.
- Gravel (2 mm < d < 20 mm): Well- or poorly-graded gravel, silty gravel, clay gravel.
- Sand (0.1 mm < d < 2 mm): silty sand, clay sand.
- Fine-grained soil (0.001 < d < 0.1 mm): Slightly plastic silt, slightly plastic clay, very plastic silt, very plastic clay.
- Organic soil.

Railway Subgrade

3.3. Hydrogeological conditions

Another fundamental parameter applied in determining the subgrade quality is hydrogeological conditions.

The various networks have tried to determine, during climatic changes, the maximum groundwater level, beyond which hydrogeological conditions are considered bad. Figure 3.2 illustrates the minimum distances of the groundwater level from a certain reference level, for hydrogeological conditions to be considered good, according to the regulations of various networks (65), (66), (67).

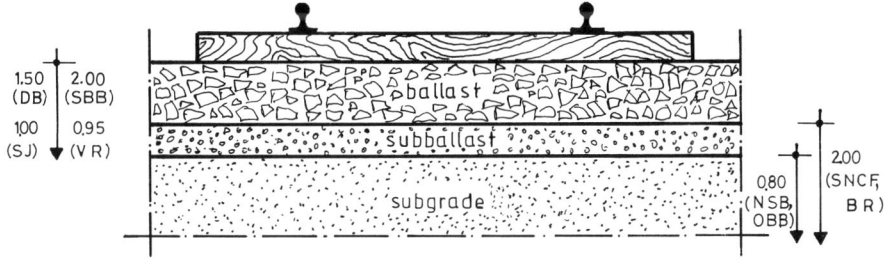

Abbreviations: BR: British Railways, DB: German Railways, NSB: Norwegian Railways, OBB: Austrian Railways, SBB: Swiss Railways, SJ: Swedish Railways, SNCF: French Railways, VR: Finnish Railways.

Fig. 3.2. Minimum distance (in metres) of the groundwater level from a certain reference level, so as drainage conditions be considered good, according to the regulations of various railway networks

Even if the groundwater level is below that shown in Fig. 3.2, however, hydrogeological conditions are not generally considered good if suitable drainage devices are not provided (Fig. 3.3) or the subballast does not have the required transverse gradient $(3 \div 5\%)$, (65), (66).

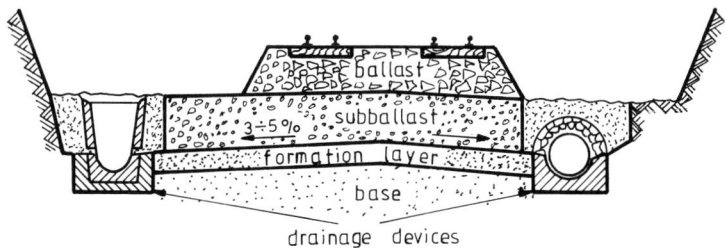

Fig. 3.3. Installation of drainage devices along the railway subgrade

Railway Engineering

Moreover, areas with large groundwater level fluctuations over time should be the subject of a separate study. In such cases, it is of interest to examine, from a technical and financial point of view, the feasibility of installing a sand filter or geotextiles. This is a subject matter referred to at the end of the chapter (section 3.13).

For northern European networks, where frost occurs frequently, a third parameter to be taken into account involves the susceptibility of the subgrade to the penetration of frost (see section 3.10 below).

3.4. Classification of the railway subgrade

In accordance with the UIC classification, the mechanical behaviour of the subgrade may macroscopically be characterized by the following:
- Low settlements and very good support of train loads. This subgrade is hereafter designated as S_3.
- Generally medium behaviour in settlements and in withstanding train loads. This subgrade is designated as S_2.
- Large settlements and less satisfactory support of loads. This subgrade is designated as S_1.
- Extensive settlements and poor performance in withstanding loads. The quality of such subgrades is designated as S_0.

To the above should be added the case of a subgrade composed of rock of satisfactory strength. The quality of such subgrade is designated as R.

The criteria for the classification into one of the above categories are geotechnical characteristics of the soil and hydrogeological conditions. Therefore, according to the applicable UIC standards, (65), the railway subgrade classification is shown in Table 3.1. The reference parameters used in this classification include the percentage of fine aggregates, plasticity index PI and the Los Angeles and Deval coefficients (see also chapter 7, section 7.4).

Soils of category S_0 are in principle unsuitable to support the track properly, because they settle extensively, they are inhomogeneous, their characteristics may change over time and, finally, they allow penetration of ballast stones deeply into the subgrade. Organic soils should be avoided whenever possible when laying out the track, or replaced by more appropriate soil material. Should this prove impossible and the track have to traverse organic soil regions, especially on high earthbanks, the risk of settlements should be considered carefully and soil improvement solutions examined in combination with the suitable increase in the ballast and subballast thickness and the use of geotextiles, (73), (74), (75).

Table 3.1.
Classification of subgrade quality as a function of geotechnical characteristics and hydrogeological conditions, (65)

Geotechnical classification of soils	Hydro-geological conditions	Railway subgrade quality
Low-variability rock	–	R
Medium-variability rock (dry Deval > 9, Los Angeles ≤ 30) Soils with fine grains* < 5%	–	S_3
High-variability rock (6 < dry Deval < 9, 30 < Los Angeles < 33) Sand with uniform fine grains < 5% Soils with fine grains 5 ÷ 15%	good bad	S_3 S_2
Schist with PI > 7 Silty sand with PI > 7 Soils with fine grains 15 ÷ 40% Crushed stone with Deval < 6 and Los Angeles > 33	good bad	S_2 S_1
Silt slightly plastic Soils with fine grains > 40%	–	S_1
Organic soils	–	S_0

* Grains are characterized as fine when their dimensions are < 60 μ.

3.5. Mechanical characteristics of the subgrade

The role of the subgrade is to withstand train loads which have been adequately attenuated by the intermediate track support structures (ballast and subballast). In order to withstand loads properly, the subgrade should have the required mechanical properties.

On the basis of a series of tests conducted within the ORE* framework, (55), the limits within which the modulus of elasticity ranges, were determined for each of the soil categories according to the UIC classification (Fig. 3.4). For rocky soils, the modulus of elasticity varies in accordance to the nature of the rock material. For the R subgrade, the modulus of elasticity is in the order of $3 \cdot 10^4$ kp/cm^2 (see also chapter 4, section 4.3.5, table 4.2).

---

* Former name of the Research Department of the International Union of Railways.

Railway Engineering

In addition to the modulus of elasticity, characterization of the subgrade also requires the determination of its carrying capacity. Figure 3.4 illustrates the respective values of the CBR index corresponding to the various subgrade soil categories, (65).

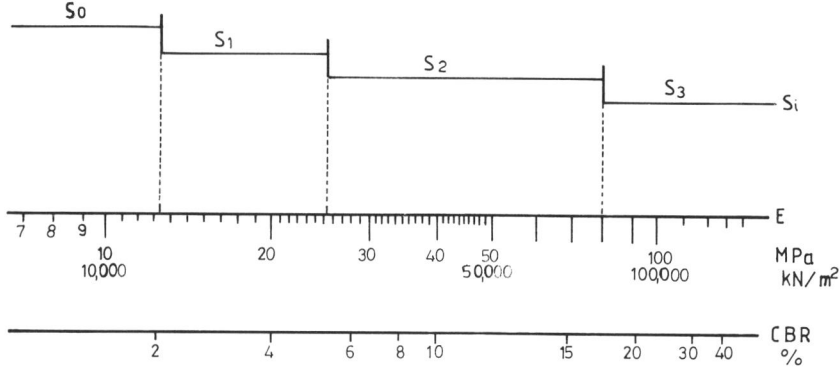

Fig. 3.4. Modulus of elasticity and CBR index for various subgrade soil categories, (65)

3.6. The formation layer

If the subgrade soil used is classified as S_1 or S_2, it is advisable to place a top layer composed of a better-quality soil material. This layer is often termed the formation layer.

The formation layer should be more compact than the base. Accordingly, most networks now require the formation layer to have a coefficient of 100% by the normal Proctor test, while this value is routinely 95% for base layers (in the case of embankments), (54).

Use of the formation layer leads to a substantial improvement in the subgrade behaviour only if the following two requirements are met, (54):
- The base material should have a low water content, otherwise the base soil particles will penetrate the formation layer, destroying the transverse slope.
- The formation layer should be homogeneous and free of local concentrations of fine-grained material.

The thickness of the formation layer is defined as a function of the subgrade quality. The values of Table 3.2 were found semi-empirically, (65).

Subgrade quality	Formation layer Quality	Formation layer Thickness (cm)
S_1	S_2	30÷55
	S_3	20÷40
S_2	S_3	20÷30

Table 3.2.
Required thickness of the formation layer as a function of the subgrade base quality for UIC 1-4 group lines, (65)

3.7. Impact of traffic load on the subgrade

When studying the impact of traffic load (line tonnage) and maintenance conditions, Dormon's rule, established in highway engineering, can be used with sufficient accuracy. According to Dormon's rule, (54), the mechanical loading of the subgrade is inversely proportional to the number of the loading cycles, raised to a power λ:

$$\frac{\sigma_1}{\sigma_2} = \left(\frac{N_2}{N_1}\right)^\lambda \tag{3.1}$$

where σ_1, σ_2 are the stresses corresponding to N_1, N_2 loading cycles, respectively, and λ is an exponent with a mean value of 0.2.

Let P be the load per axle and T the daily traffic load (tonnage) (see chapter 2, section 2.5.2). From equation (3.1) it follows that

$$\frac{\sigma_1}{\sigma_2} = \left(\frac{T_2/P_2}{T_1/P_1}\right)^\lambda \tag{3.2}$$

In the event of a constant load per axle, $P_1 = P_2$, then equation (3.2) becomes

$$\frac{\sigma_1}{\sigma_2} = \left(\frac{T_2}{T_1}\right)^\lambda \tag{3.3}$$

3.8. Impact of maintenance conditions on the subgrade

In order to estimate the magnitude (and therefore the expense) of track maintenance works, the maintenance coefficient k is used as a parameter. The entire network is divided into sections with approximately the same number of work sessions of track teams along each section, work sessions being understood to mean all sessions by manual power or by mechanical equipment between two complete renewals of the track. Let I be the annual number of work sessions along a section and I_m the average number of work sessions along tracks of the same age (i.e. renewed in the same year), belonging to the same UIC group and carrying trains with the same load per axle. The maintenance coefficient k is defined as:

$$k = \frac{I}{I_m} \quad (3.4)$$

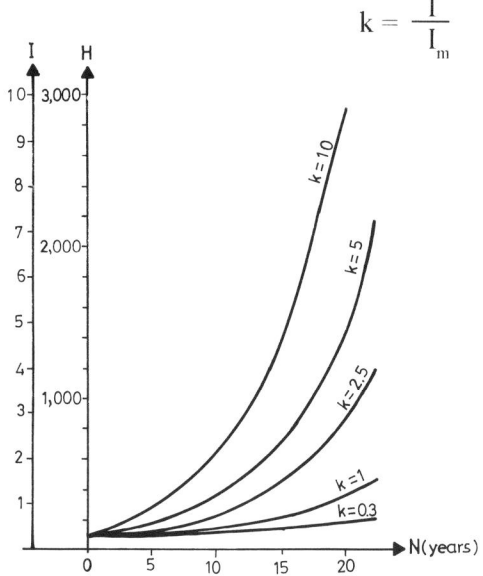

Fig. 3.5. Maintenance expenses for manual work sessions (in man-hours H per km of line) and annual number of work sessions I (both manual and by mechanical means) as a function of the maintenance coefficient k and the number of years since the last complete renewal. Case of UIC group 1-3 lines

The value k=1 corresponds to an average maintenance level, whereas the value k=0.5 corresponds to a satisfactory maintenance level. It should be noted that when subgrade quality is poor, k may take values up to 10 (Fig. 3.5).

Use of the coefficient k is of considerable assistance in the rational scheduling of track maintenance works. Figure 3.5 gives maintenance expenses (for UIC 1-3 lines) as a function of the maintenance coefficient and the number of years elapsed since the last complete renewal. On the basis of the point on the curves beyond which maintenance expenses increase disproportionately, the time at which the next complete renewal of the track should take place is rationally determined.

Let us now consider two tracks 1 and 2 with different maintenance coefficients k_1 and k_2, respectively. Application of the Dormon rule gives

$$\frac{\sigma_1}{\sigma_2} = \left(\frac{\tau_2/P_2}{\tau_1/P_1}\right)^\lambda \quad (3.5)$$

where τ is the traffic load on each line between two consecutive maintenance sessions. Statistical analysis has shown that τ is proportional to the value of $\frac{T}{k}$, therefore

$$\frac{\sigma_1}{\sigma_2} = \left(\frac{T_2/k_2/P_2}{T_1/k_1/P_1}\right)^\lambda \tag{3.6}$$

Considering the case of two lines with the same per-axle load and the same traffic load, equation (3.6) becomes:

$$\frac{\sigma_1}{\sigma_2} = \left(\frac{k_2}{k_1}\right)^\lambda \tag{3.7}$$

Equation (3.7) allows calculation of the impact of maintenance conditions on the mechanical strength of the subgrade.

The use of coefficient k requires the recording at regular intervals of the different problems arising (Table 3.3).

Table 3.3.
Record of geotechnical track problems

KILOM.		STATIONS		TRACK PROBLEMS					SECTION
FROM	TO	FROM	TO	TALUS INSTABILITY	TRACK SETTLEMENT	TRACK DISTORTION	EXCESSIVE MAINTENANCE	WATER AFFLICTED POINTS	COMMENTS
0	5	A	B						
5	10	B	C						
10	15	C	D						
...		...							

3.9. Fatigue behaviour of the subgrade

Fatigue is usually defined as the reduction of the mechanical strength of a material under the influence of repeated loads. In the case of metals, it has been found that there is a limiting stress σ_o (called fatigue limit), which if exceeded by the stresses developed, fatigue effects occur; they may lead to failure without being preceded by any macroscopically large deformations.

However, in soil materials fatigue involves the development of plastic deformations in relation to the loading cycles. Experimental results of the triaxial test under repeated loading conditions show that the parameter

$$R = \frac{(\sigma_1 - \sigma_3) \text{ of the 1st cycle}}{(\sigma_1 - \sigma_3) \text{ of the cycle causing failure}} \tag{3.8}$$

shows a limiting value on the order of 0.9, beyond which plastic deformations increase very rapidly, as apparent from Fig. 3.6.

Railway Engineering

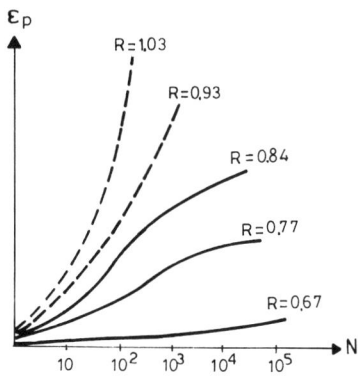

Fig. 3.6. **Evolution of plastic deformations in clay soils as a function of the R parameter.**

For the evolution of plastic deformations ε_p^N as a function of the loading cycle N, the following relation, (72), has been suggested:

$$\varepsilon_p^N = a + b \log N + cN^\alpha + dN^\beta + ... \quad (3.9)$$

where $a < b < $ .... and the parameters a, b, c, d, α, β, are determined experimentally.

According to equation (3.9), as long as the exponential terms are negligible, plastic deformation proceeds logarithmically and practically stabilizes after a certain number of loading cycles. On the contrary, if the exponential terms of equation (3.9) have a determining influence on total plastic deformation, then the subgrade may show large and dangerously increasing deformations as a function of the loading cycles. Such behaviour was observed, under certain conditions, in cases of bad (S_1) and very bad (S_0) subgrade quality.

3.10. Frost protection of railway lines

3.10.1. Frost index

The railway administration must decide whether the frost protection should be calculated according to the coldest winter possible, or whether to install a subgrade which would be suitable for average winters while accepting that frost penetration would occur in extreme conditions. Table 3.4 gives the frost index in relation to the probability of freezing through as well as the expected underratings of frost penetration in a certain period, (68).

Table 3.4.
Frost index, probability of freezing through and expected number of underratings in a certain period, (68)

Frost index	Probability of freezing through	Expected number of underratings in a certain period
F_2	50%	1 in a 2-year period
F_5	20%	1 in a 5-year period
F_{10}	10%	1 in a 10-year period
F_{100}	1%	1 in a 100-year period

Railway Subgrade

3.10.2. Frost foundation thickness

A layer of material or combination of materials is placed under the ballast layer (or the subballast) in order to protect the subgrade against frost heave. Frost foundation is a term comprising several kinds of frost-heaving prevention materials and measures.

Various materials, such as gravel, cinders, etc., can be used in the frost foundation layer. Figure 3.7 illustrates in relation to the frost index the appropriate thickness of the frost foundation layer under the ballast and Figure 3.8 illustrates the appropriate thickness when an insulation layer of foam plastic is used, (68).

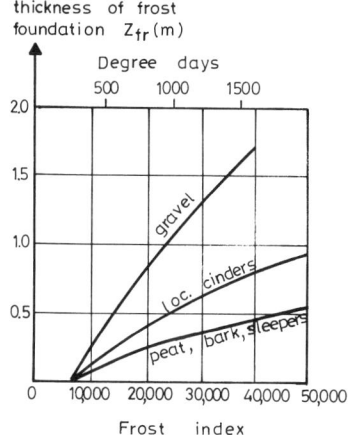
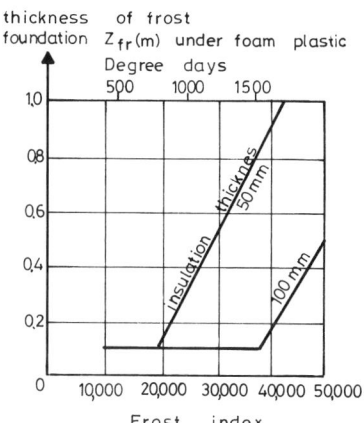

Fig. 3.7. Thickness z_{fr} of the frost foundation layer under a ballast layer of 35 cm, (68)

Fig. 3.8. Thickness z_{fr} of the frost foundation layer under a ballast layer of 25 cm, when an insulation layer of foam plastic is used, (68)

3.10.3. Frost protection methods on existing tracks

Along existing railway tracks, which cross areas often freezing in winter, many ways of improving the subgrade (during track renewal) so as to prevent against frost have been suggested, (Figures 3.9 to 3.12), (68).

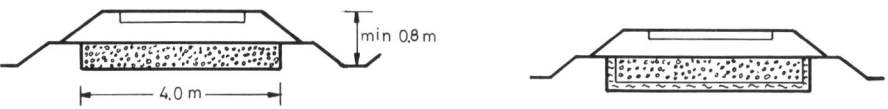

Fig. 3.9. **Frost foundation of gravel or cinders**

Fig. 3.10. **Frost foundation of stone with peat filter**

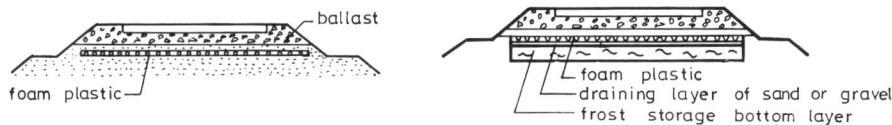

Fig. 3.11. **Frost protection by use of foam plastic**

Fig. 3.12. **Combination of insulation and a frost-storage bottom layer**

3.11. Track subgrade in trenches and on embankments - slope gradients

3.11.1. Subgrade in trench

Before excavating any trench, particular attention is made to studying the geologic formations in its path (especially in the case of diaclases), in order to disturb the geologic formation equilibrium as little as possible. Talus slides are often attributable to such oversights. Parameters to be considered when designing a trench include safety, cost, and the adaptation to the aesthetics of the surrounding environment (and not the other way around).

The slopes of the trench sides are determined from the geotechnical study, with commonly used values as follows (Fig. 3.13):

silt	1 : 2.5÷3
muddy sand	1 : 2÷3
gravel	1 : 1.5÷2
stable clay	1 : 1÷1.5

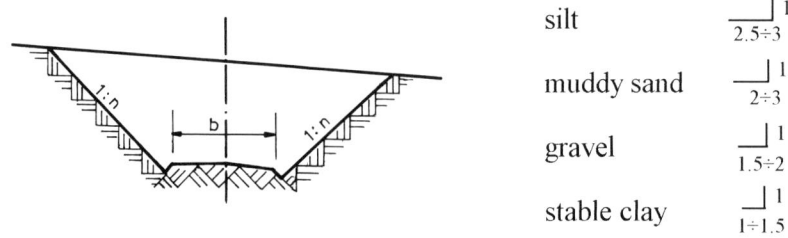

Fig. 3.13. **Slopes of trench sides in clay soil**

Railway Subgrade

Protection by talus stabilization is usually attained by covering the slopes with shrubs or by planting trees, thus at the same time achieving the merging of the work with the surrounding landscape. Ground drainage is also required to the slopes to avoid softening.

3.11.2. Subgrade on embankment

In the case of an embankment, the quality of geologic formations under the planned embankment should be also considered. Commonly used values of slope gradients are:

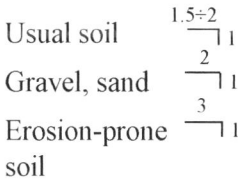
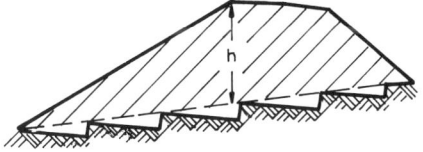

Usual soil $\quad \dfrac{1.5 \div 2}{1}$

Gravel, sand $\quad \dfrac{2}{1}$

Erosion-prone soil $\quad \dfrac{3}{1}$

Fig. 3.14. Stepping of the base of the embankment in the case of steep ground

If the ground slope is greater than 1:10, it is advisable to secure the embankment base by using a step-like configuration as shown in Figure 3.14.

Due to the subsequent compaction of the embankment, its initial dimensions should be augmented both in width and in height (Fig. 3.15).

Finally, in the case of very tall embankment sides, a retaining wall or reinforced soil designed to withstand the soil thrust and train loads may be used, (Fig. 3.16).

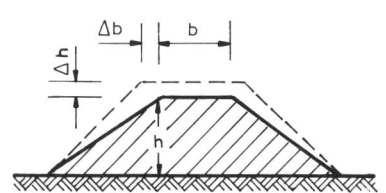

Fig. 3.15. Increase of the initial width and height due to the expected reduction in size by compaction

Fig. 3.16. Retaining wall in the case of very tall embankment sides

3.12. The reinforced soil technique in railway engineering

Reinforced soil, (78), is a flexible technique which can, in many instances, replace retaining walls. Reinforced soil is an assembly consisting of (Fig. 3.17):

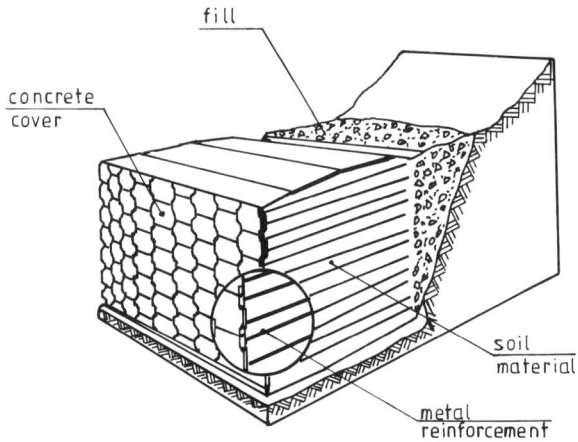

Fig. 3.17. The reinforced soil technique

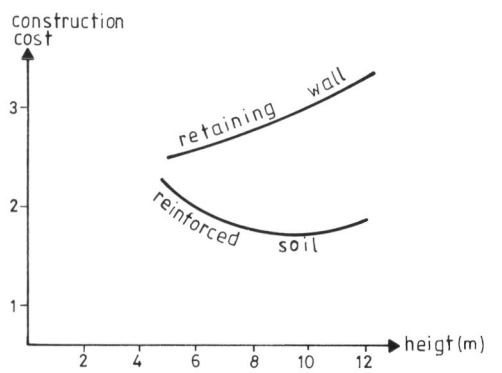

Fig. 3.18. Comparative construction cost of a retaining wall and reinforced soil in railway projects in France, (79)

- the embankment edge
- good quality soil material
- metal reinforcement
- concrete cover.

The reinforced soil technique is especially indicated in medium and poor quality subgrades (S_1, S_0). Particular attention is required in securing the metal reinforcement. A comparative analysis of the construction cost for railway projects in France (Fig. 3.18) has shown that the reinforced soil solution is both economically and technically advantageous compared to the construction of a retaining wall, especially for heights 3 m < h < 12 m. The reinforced soil technique, however, cannot be used in electric-traction lines, because the electric current return corrodes the reinforcement, which leads to failure, (78).

Railway Subgrade

3.13. Geotextiles in railway subgrades

Railway subgrades of medium, bad, or (particularly) very bad quality, can be improved through the use of geotextiles. Geotextiles, (74), (75), are permeable geomembranes consisting of synthetic polypropylene or polyester fibres, $0.4 \div 3$ mm thick, weighing $70 \div 350$ g/m length. There are two large geotextile classes:
- Woven geotextiles, composed of two interwoven perpendicular fibre layers. They are strongly anisotropic.
- Non-woven geotextiles with isotropic behaviour, in which the fibres are laid randomly.

Geotextiles have a large deformability and are used:
- to separate two consecutive layers of granular materials,
- to reinforce a soil layer of insufficient mechanical strength,
- as filters,
- for drainage.

Geotextiles are extensively used in railway engineering (Photo 3.1). They are laid under the subballast (never under the ballast) and their purpose is manifold, (73):

Photo 3.1. Positioning of the geotextile in the railway subgrade

i) *To facilitate proper and convenient laying of the track support structures on the subgrade.* The geotextile laid on top of the subgrade prevents the intrusion of fine-grained elements into the gravel subballast and allows a suitable transverse slope to be imparted to the subgrade surface. Fig. 3.19 illustrates the plasticity characteristics of certain clay soils, in the case of which a strong absorption of fine-grained materials into the superposed gravel layer was observed, (69).

ii) *To increase* (under repeated loading) *the mechanical resistance of the track support structures.* Use of geotextiles, however, should not entail an appreciable reduction of the ballast and subballast thickness, because this would result in strong loading of the subgrade, (73). Geotextiles cannot replace the ballast and gravel in distributing vertical loads. The application, by certain networks, of geotextiles

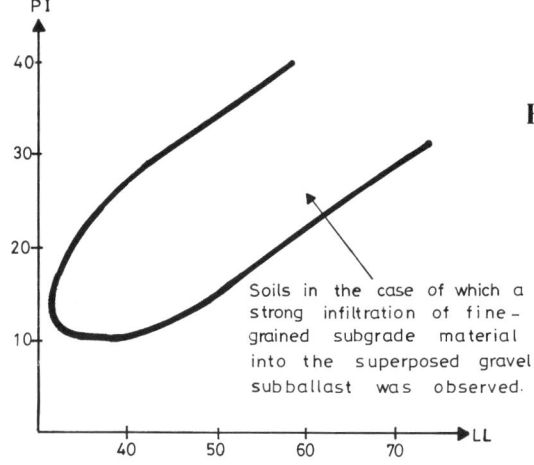

Fig. 3.19. The combination of plasticity index (PI) and liquidity limit (LL) at which a strong penetration of fine-grained subgrade elements into the gravel subballast is observed, (69)

without the laying of subballast in addition to the reduction of the ballast thickness has caused failures (perforation of the geotextile by the ballast, ruining of the transverse slope, etc.). The reinforcing action of geotextiles may be determined by numerical methods, such as finite-element analysis (see chapter 4, section 4.3), (80).

iii) *They function as filters or as drains*

In this case, the geotextile is selected from the relations:

a) Non-cohesive soils:
$$k_g \geq \frac{t_g \cdot k_s}{5 \, d_{50}} \quad (3.10)$$

b) Cohesive soils:
$$k_g > 100 \cdot k_s \quad (3.11)$$

where
- k_g : required geotextile permeability (cm/sec)
- t_g : geotextile thickness (mm)
- k_s : soil permeability (cm/sec)
- d_{50} : sieve diameter allowing passage of 50% of the soil material

Geotextiles can also protect the subgrade against frost intrusion. Before use, it should be ascertained that the specific geotextile fulfils the mechanical strength requirements: fracture strength, elongation at failure, perforation strength, compressive strength, water permeability, permeability to fine-grained soil material, etc. The values of these mechanical properties are determined by various tests described in related manuals, (74), (75).

Railway Subgrade

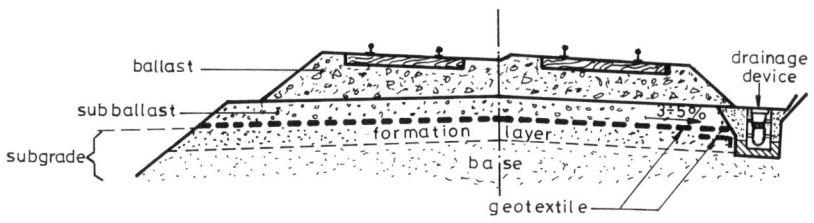

Fig. 3.20. Proper laying of geotextile on railway subgrade

Use of geotextiles along the railway subgrade usually fulfils all three purposes above. Geotextiles, however, are commonly used simply to separate the gravel subballast from the subgrade soil material.

Whenever geotextiles have been used (Fig. 3.20), track maintenance expenses have been considerably reduced. The geotextile installation expense is thus amortized very quickly.

4 Mechanical Behaviour of Track

4.1. Analysis of vertical phenomena - track coefficients

4.1.1. Definitions - symbols

We will first examine the static approach to the mechanical behaviour of the track. Let:

z: vertical settlement at the rail level
r: wheel load uniformly distributed along the rail
R: sleeper-rail vertical reaction
ℓ: sleeper spacing
S: sleeper seating area
p: average pressure applied at the sleeper seating surface on the ballast.

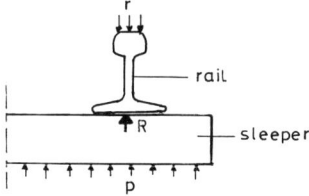

Fig. 4.1. Symbols of parameters of the track system

4.1.2. Track coefficients

We define the following track coefficients:

$$\text{Track index} \quad k = \frac{r}{z} \tag{4.1}$$

$$\text{Track stiffness} \quad h = \frac{Q}{z} \tag{4.1a}$$

$$\text{Sleeper reaction coefficient} \quad \rho = \frac{R}{z} \tag{4.2}$$

Mechanical Behaviour of Track

Substituting equation (4.1) in (4.2) we obtain

$$\rho = R\frac{k}{r} \quad (4.3)$$

and since $R = \ell r$ (equilibrium's equation), then

$$\rho = \ell r \frac{k}{r} = k\ell \quad (4.4)$$

Ballast coefficient $C = \dfrac{\rho}{S}$ \quad (4.5)

Substituting equation (4.2) in (4.5) we obtain

$$C = \frac{R}{zS} \quad (4.6)$$

and since $\dfrac{R}{S} = p$ we will have

$$C = \frac{p}{z} \quad (4.7)$$

In general terms, the *reaction coefficient* of a component of the track system is defined as

$$\rho_n = \frac{R}{z_n} \quad (4.8)$$

where z_n is the vertical settlement at the level of the examined component. Hence,

$$\Sigma z_n = z \Rightarrow \Sigma \frac{R}{\rho_n} = R\Sigma \frac{1}{\rho_n} \Rightarrow \frac{1}{\rho} = \Sigma \frac{1}{\rho_n} \quad (4.9)$$

Equation (4.9) gives the total reaction coefficient of the track-subgrade multilayer system.

Below are given values of the reaction coefficient ρ for the various track components, (56), (1):

Rail	5,000 ÷ 10,000 t/mm
Timber sleeper	50 ÷ 80 t/mm
Concrete sleeper	1,200 ÷ 1,500 t/mm
Ballast	10 ÷ 30 t/mm
Rubber pad	10 ÷ 20 t/mm

Track elasticity mainly depends on ballast thickness and characteristics. It was found, (1), that along existing tracks with only a ballast layer (i.e. with no gravel layer), the total reaction coefficient ranges between 0.15 and 1.0 t/mm, with 0.3 t/mm as an average value.

Subgrade elasticity depends on soil quality, with the following reaction coefficient values, (1):

Swampy subgrade	$0.5 \div 1.5$ t/mm
Clay subgrade	$1.5 \div 2$ t/mm
Gravel or rocky subgrade	$2 \div 8$ t/mm
Frozen subgrade	$8 \div 10$ t/mm

In civil engineering projects, the reaction coefficient values range from 10 to 15 t/mm, and therefore elasticity is far lower than in a conventional track. Rubber pads used in these cases are significantly thicker.

The increase in ballast thickness has a favourable effect not only by increasing track elasticity, but also by reducing the stresses developed in the subgrade. Let:

$$\lambda = \frac{\text{stress at the subgrade surface}}{\text{stress under the sleeper}}$$

e : thickness of the ballast layer
ρ_o: track reaction coefficient for e=0

When applying Boussinesq's analysis (multilayer system with elastic behaviour) the following values are derived (Table 4.1), (56):

Table 4.1.
Influence of ballast thickness on track elasticity and on the reduction of subgrade stress, (56)

e (cm)	0	15	20	30	40	50
λ	1	0.70	0.50	0.35	0.25	0.20
$\dfrac{\rho}{\rho_o}$	1	1.4	2.00	2.85	4.00	5.00

A detailed analysis of the influence of ballast thickness on track and subgrade stress and strain is given in section 4.3.6.

4.2. Approached elastic analysis of vertical effects - Zimmermann's method

We assume that the rail is of infinite length* and rests on a horizontal elastic layer with track index k (Fig. 4.2). The wheel load is simulated by a concen-

---

* This assumption is very nearly realized with continuous-welded rails, see chapter 5, section 5.11.

Mechanical Behaviour of Track

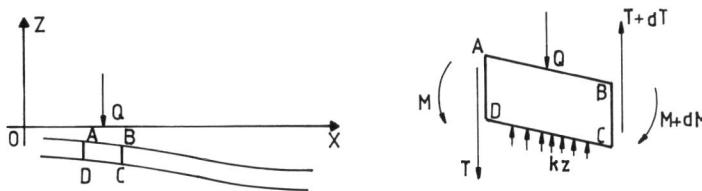

Fig. 4.2. Track simulation after Zimmermann (rail of infinite length on elastic layer) and stresses in an elementary section ABCD

trated load Q. This analysis is named after Zimmermann, (81a). The following symbols will be used:

M : bending moment
T : shear force
k : track index
E : elasticity modulus of rail
I : moment of inertia of rail

We will start by conventional strength of materials relations:

$$\frac{dM}{dx} = T \tag{4.10}$$

$$\frac{dT}{dx} = kz + Q\,\delta(x) \tag{4.11}$$

where $\delta(x)$ is the Dirac function, the Fourier transform of which is equal to one, (108).
Therefore,

$$\delta(x) = 0, \quad x \neq 0 \tag{4.12}$$
$$\delta(0) = \infty$$

The elastic line equation is:

$$\frac{d^2z}{dx^2} = -\frac{M}{EI} \tag{4.13}$$

Substituting equations (4.10) and (4.11) in (4.13) it is derived that:

$$EI\frac{d^4z}{dx^4} + kz = -Q\,\delta(x) \tag{4.14}$$

Let $Z(\omega)$ be the Fourier transform of z, and let

$$\frac{k}{EI} = w^4 \tag{4.15}$$

Equation (4.14) is transformed as*

$$\omega^4 Z + w^4 Z = -\frac{Q}{EI} \quad (4.16)$$

and

$$Z = -\frac{Q}{EI(\omega^4 + w^4)} \quad (4.17)$$

Applying the inverse Fourier tansform, it is derived that:

For $x \geq 0$, $\quad z_1 = z_o \sqrt{2}\, e^{\left(-\frac{\omega x}{\sqrt{2}}\right)} \cos\left(\frac{wx}{\sqrt{2}} - \frac{\pi}{4}\right)$ \quad (4.18)

$x < 0$, $\quad z_2 = z_o \sqrt{2}\, e^{\left(-\frac{\omega x}{\sqrt{2}}\right)} \cos\left(\frac{wx}{\sqrt{2}} + \frac{\pi}{4}\right)$ \quad (4.19)

and

$$z_{max} = z_o = \frac{Q}{2\sqrt{2}\sqrt[4]{E\,Ik^3}} \quad (4.20)$$

Therefore the analytical expressions of the bending moment, the shear force and the ballast reaction result in:

$$M = \frac{Q}{2w}\, e^{\left(-\frac{wx}{\sqrt{2}}\right)} \cos\left(\frac{wx}{\sqrt{2}} + \frac{\pi}{4}\right) \quad (4.21)$$

$$T = -\frac{Q}{2}\, e^{\left(-\frac{wx}{\sqrt{2}}\right)} \cos\frac{wx}{\sqrt{2}} \quad (4.22)$$

$$r = \frac{Q}{2}\, w\, e^{\left(-\frac{wx}{\sqrt{2}}\right)} \cos\left(\frac{wx}{\sqrt{2}} - \frac{\pi}{4}\right) \quad (4.23)$$

The graphic representation of M, T, z, (Fig. 4.3), are sinusoidal damped curves with a wavelength of

$$\lambda = 2\sqrt{2}\,\frac{\pi}{w} \quad (4.24)$$

The amplitude of the various curves is decreasing by a damping factor equal to $e^{-\pi} = 0.0432$ between consecutive waves.

As apparent in Fig. 4.3, for $x > \frac{\lambda}{2}$ the bending moment M and the shear force T are practically zero. The influence of the wheel load Q is therefore

---

* The Fourier transform F_f of a function $f(x)$ is defined by the relation:

$$F_f = \int_{-\infty}^{+\infty} f(x)\, e^{-2i\pi\omega x}\, dx$$

Mechanical Behaviour of Track

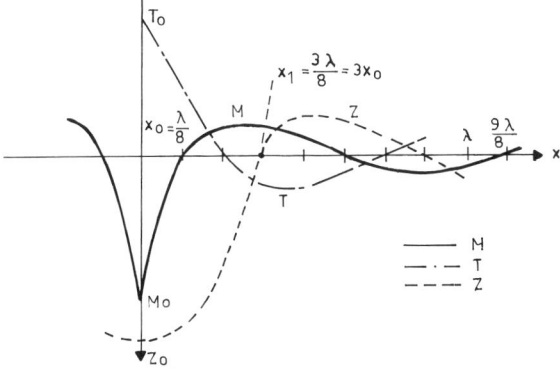

Fig. 4.3. Track stress variation as a function of distance from the point of application of the wheel load

negligible, according to Zimmermann's analysis, beyond the distance $\frac{\lambda}{2}$ from the point of application of the load Q (around 4 m).

The maximum values of the various quantities are:

$$M_{max} = M_o = \frac{Q}{2\sqrt{2}} \sqrt[4]{\frac{EI\ell}{\rho}} \qquad (4.25)$$

$$R_{max} = R_o = \frac{Q}{2\sqrt{2}} \sqrt[4]{\frac{\ell^3 \rho}{EI}} \qquad (4.26)$$

$$z_{max} = z_o = \frac{Q}{2\sqrt{2}} \sqrt[4]{\frac{\ell^3}{EI\rho^3}} \qquad (4.27)$$

$$h = \frac{Q}{z_o} = 2\sqrt{2} \sqrt[4]{\frac{EI\rho^3}{\ell^3}} \qquad (4.28)$$

Zimmermann's method is approximate, but permits convenient and rapid calculation of the maximum values of the various parameters and gives the engineer an adequate estimate of the order of magnitude of the effects under study.

Equations (4.25) to (4.28) show that if the sleeper reaction coefficient ρ increases, M_o and z_o decrease and R_o increases. The vertical settlement z_o, however, which is proportional to $1/\rho^{3\cdot4}$ decreases much faster than the bending moment M, which is proportional to $1/\rho^{1\cdot4}$. Therefore, a good value of the sleeper reaction coefficient influences more track geometry than rail mechanical behaviour. It should be noted that the sleeper reaction coefficient is mainly affected by the quality of the subgrade, where most of the total vertical settlement occurs.

Railway Engineering

An increase of sleeper spacing ℓ results in an increase of the various stresses. Vertical settlement and sleeper reaction increase faster than moment, because the former is proportional to $\ell^{3/4}$, while the moment is proportional to $\ell^{1/4}$. Consequently, a reduction of sleeper spacing affects more track geometry and less rail mechanical behaviour.

When rail stiffness EI increases, M_o increases and z_o and R_o decrease. Rail stiffness increases mainly as a result of an increase of rail weight per unit length.

As is well-known from the Strength of Materials, rail bending stresses are found from the relation

$$\sigma = M \frac{y}{I} \qquad (4.29)$$

and for the present case

$$\sigma_{max} = \frac{Q\, y_o}{2\sqrt{2}} \sqrt[4]{\frac{E\ell}{I^3 \rho}} \qquad (4.30)$$

where y_o is the maximum distance from the rail centre of gravity.

Therefore an increase of the rail moment of inertia significantly influences the stresses generated within the rail and in a lesser degree the track geometry. This is why the increase of the load per axle in recent years led to a considerable increase in the rail cross-section.

4.3. Accurate static analysis of vertical effects - Finite-element method

4.3.1. Advantages and procedure of the finite-element method

Approximative methods (Zimmermann's method, Boussinesq's multilayer method, etc.) permit a convenient and quick approximation, but their deviation from actual values, as shown by on-site measurements, may reach 100%, (55). It is therefore necessary that the mechanical behaviour of the track-subgrade system (mainly of the strain and stresses developed, on which the dimensioning of the various layers will be based) be analyzed by more accurate methods. This is now relatively easy, given the numerical methods and powerful computers. One of the methods permitting accurate analysis of the phenomena is the finite-element method, (85), (86), (91), (92), (93). According to this method, the subject under study, instead of being a physical system (Fig. 4.4.a), is a system resulting from dividing the physical system into discrete parts (finite elements). For reasons of symmetry, the study of the problem in the divided system is limited to 1/4 of the initial system (Fig. 4.4.b). The discretization of the system (construction of the mesh of the model) is an essential part of the method and the resulting finite elements must be homogeneous (i.e. of about the same size), otherwise the convergence of the method may suffer.

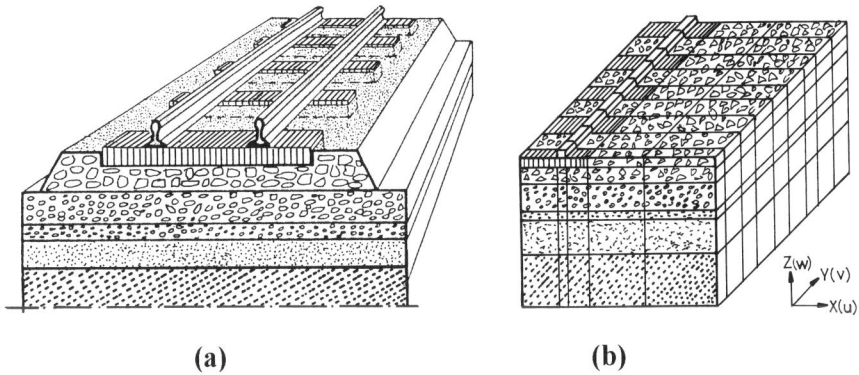

Fig. 4.4. The railway system (a) and its discretization into finite elements (b)

The finite-element method permits to study the actual physical system without coarse simplifications, to take into account exact limit conditions (i.e. the conditions imparting to the stresses or strains specific values at limit positions, e.g. in supports the displacement = 0) and the exact constitutional law of behaviour (i.e. the relationship between stresses-strains for every material). The limit conditions and the constitutional law of behaviour are also two essential parts of the finite-element method.

4.3.2. Limit conditions

The limit conditions considered are as follows:
- conditions by symmetry, i.e. transverse displacement at any plane of symmetry is zero
- conditions at the most distant points of the problem, where vertical displacement to the plane considered is set to zero.

Limit conditions must be set in such a way that the finite-element model will have a similar behaviour to the physical system, of which the real behaviour is being investigated.

4.3.3. Stress-strain relationship

The constitutional law of behaviour (stress-strain relationship) must express the real and physical behaviour of the materials. Concerning ballast and subgrade, it was found that the deformation caused by passing train loads consists of two components:

- an elastic component which disappears after passage of the train
- a plastic component remaining after the train has passed.

4.3.3.1. Case of ballast and subgrade

The behaviour of the ballast, the gravel subballast and the subgrade, as tested by in situ experiments, (55), (99), is found to be elastoplastic and is given by the following equations:

$$\varepsilon_{ij}^{total} = \varepsilon_{ij}^{elastic} + \varepsilon_{ij}^{plastic} \tag{4.31}$$

$$\varepsilon_{ij}^{elastic} = \frac{1+\nu}{E}\sigma_{ij} - \frac{\nu}{E}I_1\delta_{ij} \tag{4.32}$$

$$\varepsilon_{ij}^{plastic} = \lambda \frac{\partial f}{\partial \sigma_{ij}} \tag{4.32a}$$

where ε_{ij}^{total} : total deformation
$\varepsilon_{ij}^{elastic}$: elastic deformation
$\varepsilon_{ij}^{plastic}$: plastic deformation
E : modulus of elasticity
ν : Poisson's ratio
$I_1 = \sigma_{11} + \sigma_{22} + \sigma_{33}$
δ_{ij} : Kronecker's delta, $\delta_{ij} = 1$ for i=j, $\delta_{ij} = 0$ for i≠j
f : plasticity criterion, with a different expression for each material
λ : scalar quantity
The indices i, j take the values 1, 2, 3.

It has been proven that the plasticity criterion best suited for *soil materials* and *ballast* is the *Drucker-Prager* criterion, defined by the equation, (90):

$$f(\sigma) = \alpha I_1 + J_2 - k \tag{4.33}$$

where

$$I_1 = \sigma_{11} + \sigma_{22} + \sigma_{33} \tag{4.34}$$

$$J_2 = \frac{1}{6}[(\sigma_1 - \sigma_2)^2 + (\sigma_2 - \sigma_3)^2 + (\sigma_1 - \sigma_3)^2] \tag{4.35}$$

$\sigma_1, \sigma_2, \sigma_3$: principal stresses

$$\alpha = \frac{\tan \varphi}{(9 + 12 \tan^2\varphi)^{1/2}} \tag{4.36}$$

$$k = \frac{3c}{(9 + 12\tan^2\varphi)^{1/2}} \qquad (4.37)$$

c : cohesion
φ : friction angle

If the track support is a *concrete slab*, the plasticity criterion is best represented by the *parabolic criterion*, expressed, (88), by:

$$f(\sigma) = J_2 + \frac{1}{3}(R_c - R_T)I_1 - \frac{1}{3}R_cR_T \qquad (4.38)$$

where R_c : compressive strength
       R_T : tensile strength

4.3.3.2. Case of rail and sleeper

In contrast to ballast and subgrade, rails and sleepers show an almost elastic behaviour, i.e. plastic deformations are negligible and may be ignored. Whenever plasticity effects have to be taken into account, however, the *parabolic criterion* should be used as the plasticity criterion for *concrete sleepers*. For *rails*, the *von Mises* criterion should be used, conforming, (89), to the equation:

$$f(\sigma) = \sqrt{\frac{1}{6}[(\sigma_1 - \sigma_2)^2 + (\sigma_2 - \sigma_3)^2 + (\sigma_1 - \sigma_3)^2]} - q \qquad (4.39)$$

where q : shear elasticity limit.

4.3.4. Numerical calculation procedure

In finite-element analysis, three categories of models have been developed, (86), (87):
- *Strain* (or kinematic) models, in which the limit conditions concerning strains (deformations) are introduced as given data, and equilibrium equations as well as limit conditions concerning forces are the object of successive approximations. Strain models have proven more convenient both in synthesis and in use.
- *Stress* (or static) models, in which equilibrium equations and limit conditions concerning stresses are introduced as known data.
- *Hybrid* models, where the strain approximation is applied in a geometric part of the model and the stress approximation in another part.

In *strain models* in particular, static finite-element analysis leads to the solution of the system

$$[k][q] = [F] \tag{4.40}$$

where [k] : the system's stiffness matrix
[q] : the displacement vector of the system's nodes
[F] : the vector of the forces exerted on system nodes.

The quantities [k], [q], [F] for the entire system are the result of assembly of the elementary quantities $[k_e]$, $[q_e]$, $[F_e]$ corresponding to each finite element.

The law of elastoplastic behaviour correlating stress and strain may be approximated by two methods, (87):
  a. The initial stress method, which is slower in convergence but easier to use (Fig. 4.5.a).
  b. The variable stiffness method, which has the significant disadvantage of the need to reverse the stiffness matrix at each successive approximation (Fig. 4.5.b).

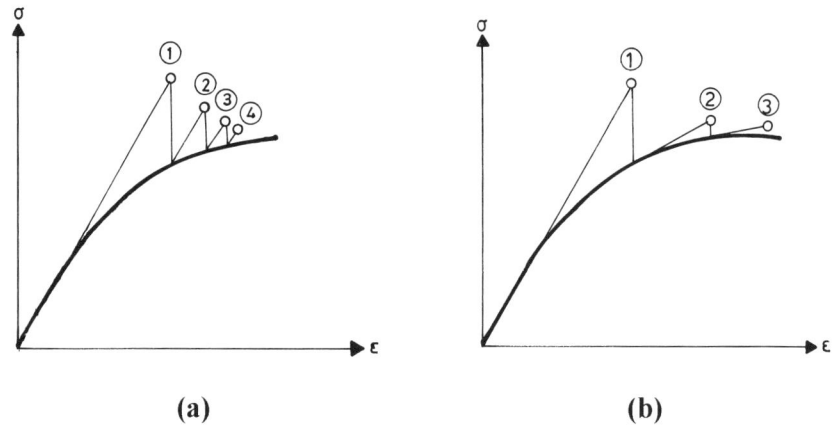

(a) (b)

Fig. 4.5. The initial stress (a) and variable stiffness (b) methods to approximate the elastoplastic stress-strain relationship

4.3.5. Determination of the mechanical characteristics of the various materials

The subgrade is assumed to be divided into various categories (S_1, S_2, S_3) (see chapter 3, section 3.4). Table 4.2 gives the average values of the mechanical characteristics, as determined by a series of tests conducted within the framework of the International Railways Union (UIC), (98), (99).

Table 4.2.
Values of the mechanical characteristics of railway track and subgrade materials, (98), (99)

Material	Elasticity modulus E (kp/cm²)	Poisson's ratio ν	Cohesion c (kp/cm²)	Friction angle $\varphi(°)$
Poor-quality subgrade (S_1)	125	0.4	0.15	10
Medium-quality subgrade (S_2)	250	0.3	0.10	20
Good-quality subgrade (S_3)	800	0.3	0	35
Rocky subgrade (R)	$3 \cdot 10^4$	0.2	15	20
Ballast	1,300	0.2	0	45
Gravel subballast	2,000	0.3	0	35
Sand	1,000	0.3	0	30

Material	Elasticity modulus E (kp/cm²)	Poisson's ratio ν	Compressive strength R_T (kp/cm²)	Tensile strength R_c (kp/cm²)
Reinforced concrete sleeper	$30 \cdot 10^4$	0.25	30	300
Prestressed concrete sleeper	$50 \cdot 10^4$	0.25	60	90
Tropical timber sleeper	$25 \cdot 10^4$	0.25	100	1,000
Rail (steel)	$2.1 \cdot 10^6$	0.30	$7 \cdot 10^3$	$6 \cdot 10^3$

4.3.6. Stress and strain in the track-subgrade system according to the finite-element method

Finite-element analysis allows all track-subgrade system parameters to be taken into consideration, (91), (92), (94):
- Subgrade soil quality (S_1, S_2, S_3) (see chapter 3, section 3.4).
- Sleeper type (see chapter 6, sections 6.3, 6.5, 6.6):
  • Twin-block reinforced-concrete sleeper
  • Monoblock prestressed-concrete sleeper
  • Timber sleeper
- Track support thickness e (= ballast + subballast) (see chapter 2, section 2.2 and chapter 7, section 7.5).

Figs. 4.6, 4.7, 4.8 that follow, illustrate the vertical stresses at the subgrade level, as well as the vertical settlements at the rail, sleeper, and subgrade level,

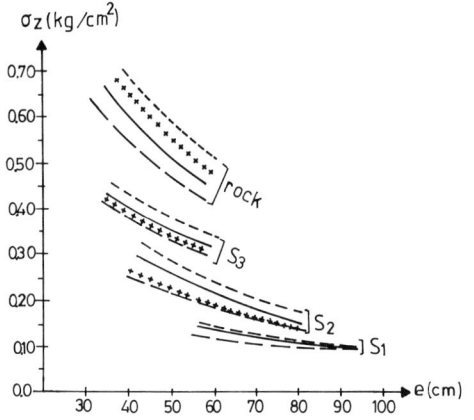

Fig. 4.6. Vertical stresses at the subgrade level for various soil and sleeper types, as a function of track support thickness e (=ballast + subballast). Elastoplastic finite-element analysis, (91), (92)

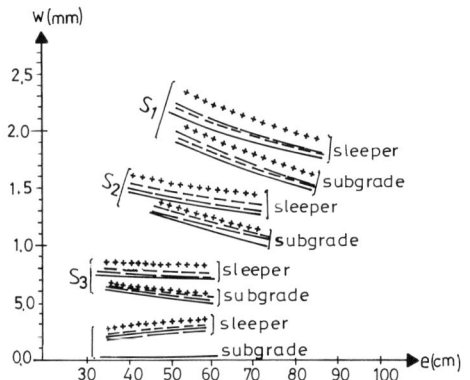

Fig. 4.7. Vertical settlements at the subgrade and sleeper level for various soil and sleeper types, as a function of track support thickness e. Elastoplastic finite-element analysis, (91), (92)

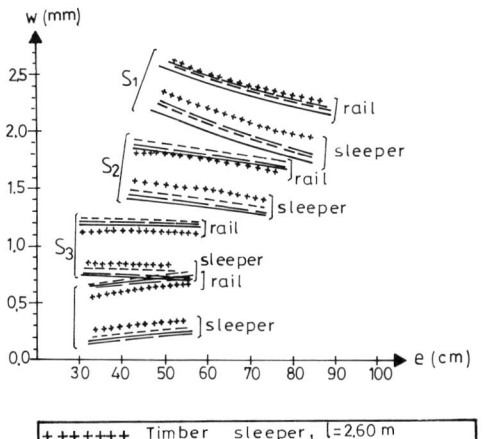

Fig. 4.8. Vertical settlements at the sleeper and rail level for various soil and sleeper types, as a function of track support thickness e. Elastoplastic finite-element analysis, (91), (92)

according to elastoplastic analysis by the finite-element method, (55), (91), (92). It is observed that the various stress quantities are primarily affected by subgrade soil quality and to a lesser degree by the track support thickness e. Indeed, the better the subgrade soil quality is, the lesser the influence of thickness e. In particular and with all other parameters unchanged, an improvement of subgrade quality from one class to the next $(S_1 \to S_2, S_2 \to S_3, S_3 \to R)$ will result in an increase of the stresses developed in the subgrade by about 50%. With respect to the influence of the sleeper type, it is observed (except in the case of a rocky subgrade) that timber sleepers and monoblock prestressed-concrete sleepers are superior in load distribution, i.e. they result in reduced stresses in the subgrade. In any case, the influence of sleeper type is again smaller than the influence of subgrade quality.

4.3.7. Distribution of wheel load along successive sleepers

The view used to prevail in railway engineering, on the basis of empirical considerations, was that when a wheel load is applied above a sleeper, then the sleeper below the load supports 50% of the wheel load and each of the neighbouring sleepers supports another 25%. Stress measurements, however, have shown that this is not the case. Finite-element analysis has shown that wheel load distribution along successive sleepers is as follows (Fig. 4.9), (142):

- Sleeper under loading wheel: 40%
- First neighbouring sleeper: 23%
- Second neighbouring sleeper: 7%.

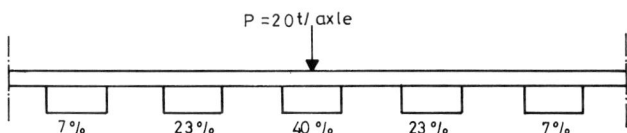

Fig. 4.9. Wheel load distribution along successive sleepers, (142)

Therefore, when a wheel load is applied over a sleeper, its effect is negligible beyond the third successive sleeper. The above load distribution in conjunction with the value of the wheel load affect sleeper dimensioning.

4.3.8. Sleeper elastic line

The elastic line is an essential part of the railway system mechanical behaviour analysis. Figure 4.10 compares the elastic line for timber and for prestressed-

Railway Engineering

concrete sleepers. Figure 4.11 illustrates the timber sleeper elastic line for various qualities of subgrade, (142). The significant role of the subgrade is again confirmed.

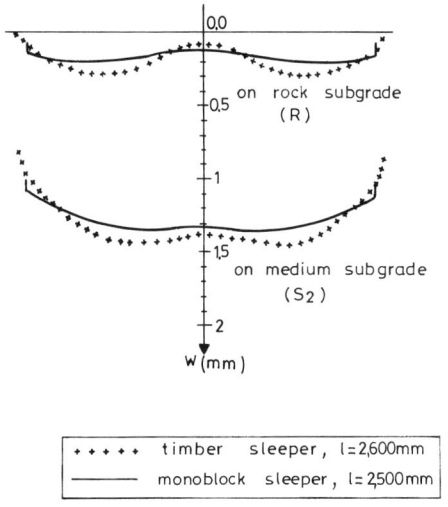
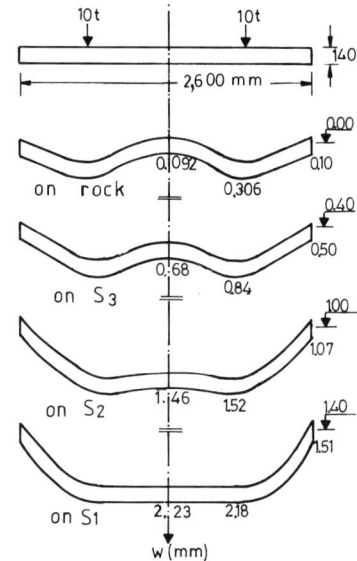

Fig. 4.10. Comparative elastic line for timber sleeper and monoblock prestressed-concrete sleeper, (142)

Fig. 4.11. Elastic line of timber sleeper for various subgrade qualities, (192)

4.4. Dynamic analysis of the track-subgrade system

As discussed in section 2.11.2 of chapter 2, an adequate approximate analysis of the stresses and strains of the track-subgrade system may be obtained by static analysis, neglecting dynamic effects, which complicate calculations. A comparison of the results of finite-element static analyses with stress and strain measurement results, has shown deviations not exceeding 20%, thus confirming that the static approach is satisfactory for stress and strain analysis, (55), (95).

There are phenomena, however, which cannot be adequately approximated by the static approach. These include the problem of the transmission of vibrations from the trains to the environment and the problem of the motion and suspension of the various rolling stock components, (103), (105).

Mechanical Behaviour of Track

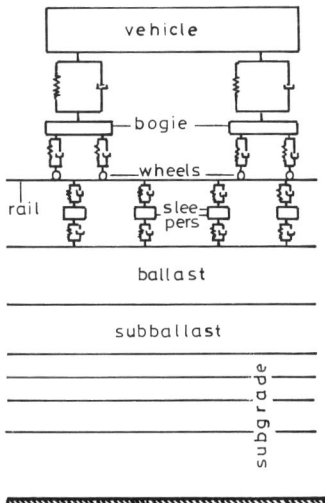

Fig. 4.12. Modelization of the rail-vehicle-track-subgrade system for a dynamic analysis

A good approximation of dynamic effects can be realized by a viscoelastic law of behaviour and is shown in Fig. 4.12, where:
* The symbol —ww— represents elastic behaviour

* The symbol —[·— represents viscous behaviour

* The symbol represents viscoelastic behaviour (Kelvin-Voigt model)
* Rail vehicles and bogies are modelized as non-deformable solids
* Wheels and sleepers are modelized as discrete masses
* The ballast and the various subgrade layers are modelized as horizontal layers
* The various system components are interconnected by a viscoelastic stress-strain relationship (Kelvin-Voigt model).

In the dynamic approach, the problem is reduced to solving the dynamic equation

$$[M][\ddot{q}] + [C][\dot{q}] + [k][q] = [F] + [R] \qquad (4.41)$$

where [M] : the mass matrix
[C] : the viscosity (damping) matrix
[k] : the stiffness matrix
[q] : the displacement vector
[q̇] : the velocity vector
[q̈] : the acceleration vector
[F] : the external forces vector
[R] : the vector of the reactions exerted by the sleeper on the ballast.

In the dynamic approach, calculations are complex and take a long time. As a result, they should be restricted only to phenomena which cannot be adequately approximated by static calculations.

4.5. Additional dynamic loading

Analysis of the mechanical behaviour of the rail vehicle has until now been based on the assumption that both rails and wheels are smooth and free of defects. However, this is not the case, and, as explained in chapter 2 (section 2.11.2), these defects excite the system and cause additional dynamic loading Q_{dyn}, which may reach values of up to 50% of the wheel load.

The mechanical loading of the track-subgrade system should therefore be considered not on the basis of the static per-wheel load Q_{stat}, but according to the total load (Fig. 4.13)

$$Q_{tot} = Q_{stat} + Q_{dyn} \qquad (4.42)$$

Additional dynamic loads may be divided into three categories according to the respective vibration frequency (Fig. 4.14):

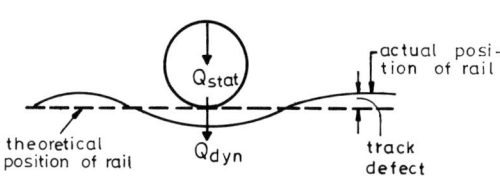

Fig. 4.13. Track defects and additional dynamic loading

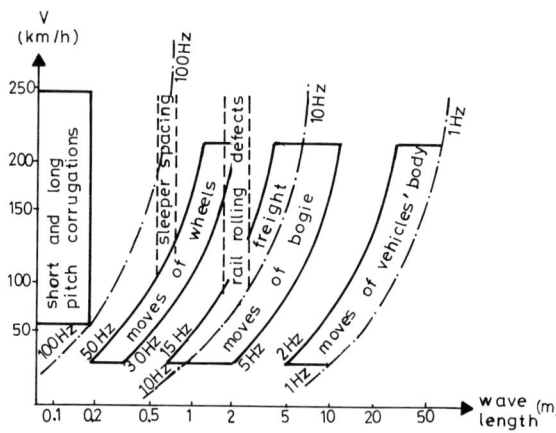

Fig. 4.14. Dynamic loading of track and corresponding frequencies, (4)

• Loads in the range $0.5\,Hz < \nu < 15\,Hz$. These correspond to the movement of sprung masses and depend principally on the characteristics and peculiarities of the rolling stock.

• Loads in the range $20\,Hz < \nu < 100\,Hz$. These correspond to the movement of unsprung masses (wheels, rails), and depend mainly on track quality and stiffness.

• Loads in the range $100\,Hz < \nu < 2,000\,Hz$. These correspond to short and long pitch corrugations of the rail surface (see also chapter 5, sections 5.7.3, 5.7.4).

Assuming a linear behaviour, (106), (107), it is possible to separate each class of additional dynamic loading from others. In order to correlate track

Mechanical Behaviour of Track

defects (see chapter 11, section 11.3) and the resulting dynamic loading Q_{dyn}, spectral analysis is used, since track defects may be recorded accurately and in detail by special recording vehicles. Analysis is again based on the dynamic equation (4.41).

The usual procedure in railway engineering is to carry out a static or pseudostatic analysis. The question, however, which arises is what is the dynamic impact factor η by which the static load should be increased in order to take into account dynamic effects. Figure 4.15 summarizes results of various theories.

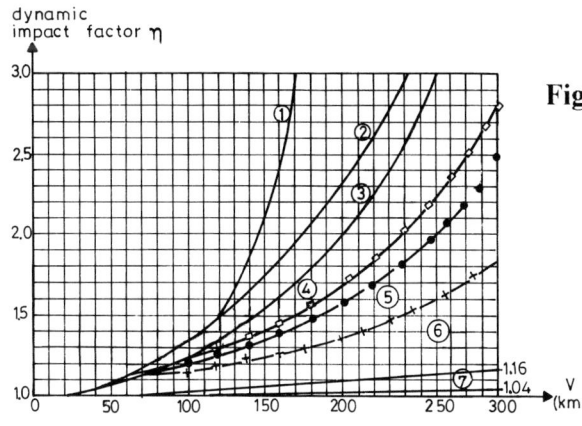

Fig. 4.15. Results for the dynamic impact factor η according to various theories, (107a)

Legend

① Winker (1871), $\eta = \dfrac{1}{1 - \dfrac{M_o V^2}{E I g}}$, $M_o = 0.1188\, P(t) \cdot \ell$ (cm), $P(t)$: axle load, ℓ (cm): sleeper spacing, I (cm⁴): moment of inertia, V (km/h)

② Formula of Central European Railways (1936), $\eta = 1 + \dfrac{V^2}{30{,}000}$

③ $\eta = \dfrac{1}{(1 \div 7) \cdot 10^{-8} V^2 \cdot \dfrac{p \cdot \ell}{I}}$ for V < 65 km/h

$\eta = \dfrac{1}{[1 - (9.1 \cdot 10^{-6} V - 2.957 \cdot 10^{-4}] \dfrac{p \cdot \ell}{I}}$ for V > 65 km/h

④ Schramm's formula (1955), $\eta = 1 + \dfrac{4.5\, V^2}{100{,}000} - \dfrac{1.5\, V^2}{10{,}000}$

⑤ Birman's formula (1966), $\eta = 1 + \alpha + \beta + \gamma$, $\alpha = 0.04 \left(\dfrac{V}{100}\right)^3$, $\beta = 0.2$, $\gamma = \gamma_o \cdot \alpha \cdot \beta$,

$\gamma_o = 0.1 + 0.017 \left(\dfrac{V}{100}\right)^3$

⑥ Values measured on vehicles of TGV 001 (1981)
⑦ Values for theoretical calculation for ideal track and vehicle without irregularities.

Railway Engineering

Figure 4.15 shows huge differences between ideal track theoretical calculation (curve 7) and measured values (curve 6) or values taking into account rail imperfections and defects (curves 1 ÷ 5). However, curves 1 ÷ 3, are deduced from old rolling stock characteristics and cannot be valid for modern rolling stock. More close to reality are curves 4, 5, 6 which show that for speeds up to 200 km/h the dynamic impact factor η varies from 1.35 to 1.6. Thus, for speeds up to 200 km/h a dymanic impact factor of 1.5 is suggested. For speeds greater than 200 km/h an analytical survey should be conducted based on experimental data.

4.6. Design of the track-subgrade system

The track-subgrade system is designed according to the following two principles, (91):
- Loads must be properly distributed to the various layers and adequately withstood by the subgrade, where the developed stresses must be less than the values causing failure

$$\sigma_{developed}^{subgrade} < \sigma_{failure}^{subgrade} \qquad (4.43)$$

- Adequate flexibility of the system should be ensured, i.e. track stiffness (see section 4.1.2) should not be excessive. Track stiffness is mainly determined by subgrade soil quality and ballast thickness. Rocky subgrades, for instance, with no problem as regards proper application of external loads, have a stiffness more than triple compared to clay subgrades. Accordingly, rocky subgrades, although free of load distribution problems, always receive a ballast + subballast layer. Finally (see chapter 2, section 2.11.2) it should be noted that additional dynamic loads increase when track stiffness increases.

4.7. Vibrations from rail traffic

4.7.1. Origin of rail vibrations

A surface vibrating source is producing three types of waves, (102):
- pressure waves (7% of the energy transmitted)
- shear waves (26% of the energy transmitted)
- Rayleigh waves (67% of the energy transmitted)

Rail vibrations have two principal origins:
- rolling engines
- wheel-rail interaction

In electrified lines, (see chapter 13, sections 13.8, 13.9), a third origin should be added, the catenary noise, caused by friction from the sliding contact of the pantograph-slider and trolley wire.

Mechanical Behaviour of Track

4.7.2. Relation of rail noise level to speed

From various analyses, (107a), is found a logarithmic relationship of the level of rail noise L to train speed V, of the form:

$$L \, (dB(A)) = A + B \log V \quad (4.44)$$

with A, B coefficients depending on rolling stock and track characteristics, type of traffic, soil characteristics, etc.

4.7.3. Damping of rail noise in relation to distance

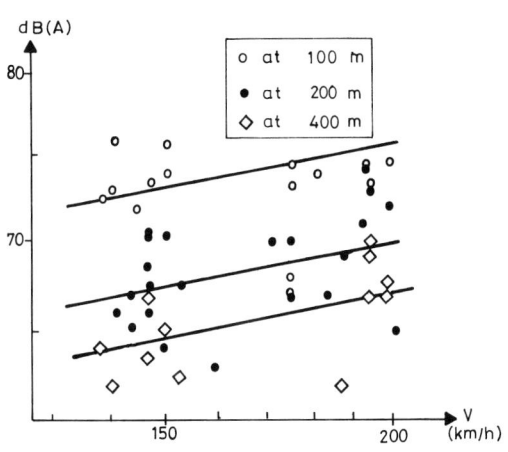

Fig. 4.16. Rail noise level in relation to distance and speed, (107a)

Figure 4.16 illustrates the noise level (in dB(A)) in various distances (100 m, 200 m, 400 m) from the rail and for speeds from 80 to 200 km/h. One can note that:
- the noise level does not decrease linearly for each doubling of distance as would be expected, probably due to ground impedance,
- noise levels are influenced more by distance than by changes in speed,
- noise levels are correlated with the logarithm of speed.

4.7.4. Noise level in relation to infrastructure type

Measurements of noise level at 25 m from the track centerline have been conducted at a speed of 200 km/h in the Japanese Shinkansen (with 12÷16 vehicles) for various infrastructure types: bridge, viaduct, embankment, cut, (Fig. 4.17).

Of note are the strong low-frequency components for the bridge and viaduct, apparently from structure radiated noise. The levels in the cut show the effectiveness of the cut in blocking the sound propagation path. Consequently, geometric design can be used as a way of reducing the impact of rail vibrations.

Railway Engineering

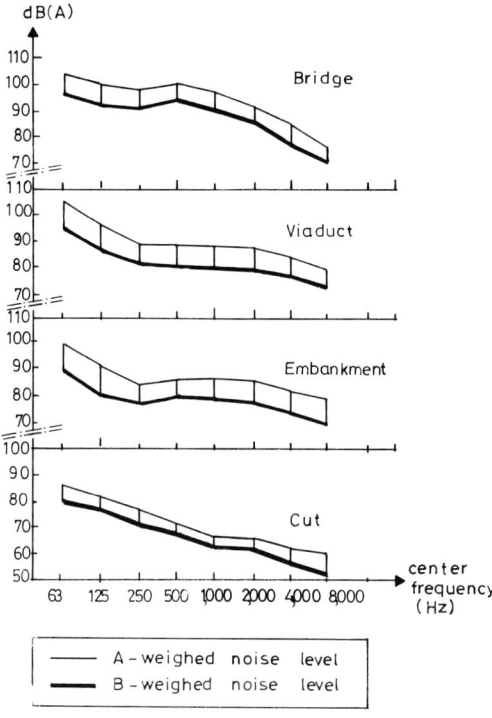

Fig. 4.17. Noise level in relation to infrastructure type, (107a)

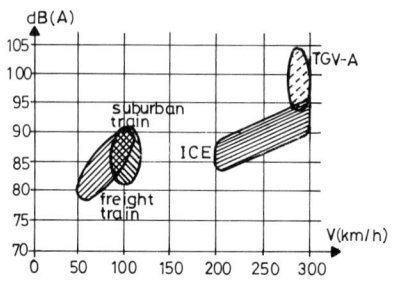

Fig. 4.18. Noise level in relation to the type of train

4.7.5. Noise levels in high speeds

A major concern in high speed trains is to reduce noise levels emitted. Thus a noise level of 97 dB(A) is reported (in 1983) for the French TGV at 25 m from the track and a speed of 272 km/h. For the German ICE, noise levels of 86 and 93 dB(A) have been reported at a distance of 25 m for speeds of 200 and 300 km/h respectively, (107a). Figure 4.18 illustrates noise levels in relation to the type of the train.

4.7.6. Noise level standards

If noise level cannot be reduced otherwise (e.g. by the appropriate design of rolling stock and track), the usual means (in order to comply with noise level standards) is to construct noise barriers along the track, so as to protect neighbouring sensitive human activities.

In recent years, national and international specifications require environmental studies in cases of important projects, such as new railways. Standards for noise level differ from one country to another.

5 The Rail

5.1. Rail profiles

Rails support and guide the wheels of the vehicle. Their profile has been the object of continuous improvement since the appearance of railways.

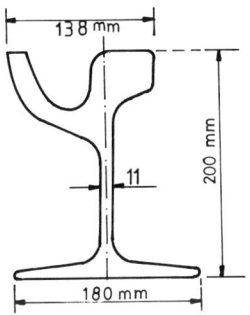

Fig. 5.1. Grooved rail

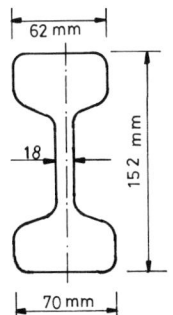

Fig. 5.2. Double-headed (or bull head) rail

Of the first rail profiles, the only one surviving to this day is the grooved rail shown in Fig. 5.1, which is presently in use along tracks where the rail top and the pavement surface are at the same level. These include tracks connecting port facilities and tramway lines.

The double-headed or bull head rail (Fig. 5.2) was widely used in the last century, with the expectation that when the upper section was worn, the rail would be reversed to use the lower portion which, according to the assumption of the time, would have remained intact. Facts did not vindicate this assumption, however, and the double-headed rail was abandoned in many countries at the beginning of the 20th century, although it is still in use on some railways and metros (U.K. and elsewhere).

The rail type which finally prevailed and is currently widely used is the rail with base (Fig. 5.3), also known as the flat bottom rail, or Vignole-type rail, named after the British engineer who

Railway Engineering

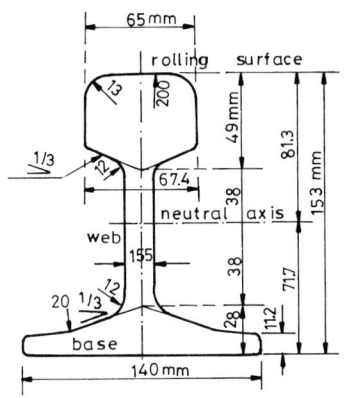

Fig. 5.3. Flat bottom or Vignole-type rail. U 36 section (50 kg/m)

designed it. This rail consists of the head, the web, and the base (Fig. 5.3). The cross-section characteristics are the weight m per unit length and the moment of inertia I. A constant goal has been to make any increases of m contingent on a proportionally greater increase of I, to ensure that the I/m ratio increases faster than m. This has led to a constant increase of the height of the rail.

The flat bottom or Vignole-type rail cross-section was formulated on the basis of the need to join rail lengths together, which can be realized with fishing plates (see section 5.10). The extensive use of continuous-welded rails (see section 5.11), however, is likely to lead in the future to a change in the rail profile.

The increase of the load per axle and of train speed has increased rail loading. The cross-sections of standard-gauge rails have been standardized by the UIC, with main types UIC 50 (weight: 50.18 kg/m), UIC 54 (weight: 54.43 kg/m), UIC 60 (weight: 60.34 kg/m), UIC 71 (weight: 71.19 kg/m). Figures 5.4 and 5.5 illustrate cross sections of rail profiles UIC 50, 54, 60 and 71.

5.2. Choice of rail section

The choice of rail section mainly depends on the traffic load (see chapter 2, section 2.5.2) as well as on the intervals between renewal sessions. For a standard-gauge track, it is customary to use UIC 50 rail for low traffic load, and UIC 60 for medium and heavy load. UIC 71 section has been recently introduced but has not been used until the mid-1990s. Table 5.1 gives the rail choice criteria as a function of traffic load, (107).

Table 5.1.
Choice of the rail section in relation to traffic load

Daily line traffic load (tons)	< 25,000 t	25,000 ÷ 35,000 t	> 35,000 t
Required weight m of the rail per metre of length	50 kg/m	◆ For timber sleepers: 50 kg/m ◆ For concrete sleepers: 60 kg/m	60 kg/m

The Rail

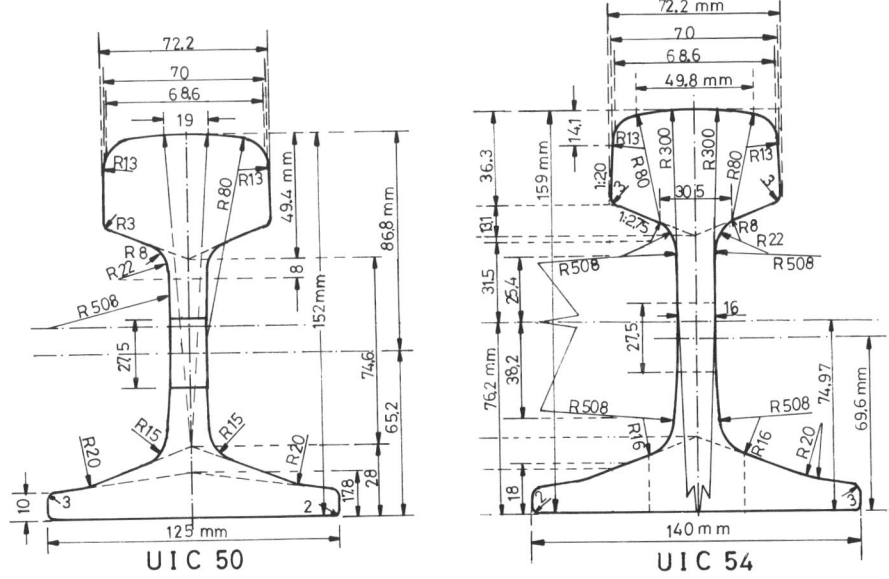

Fig. 5.4. Rail profiles UIC 50 and UIC 54

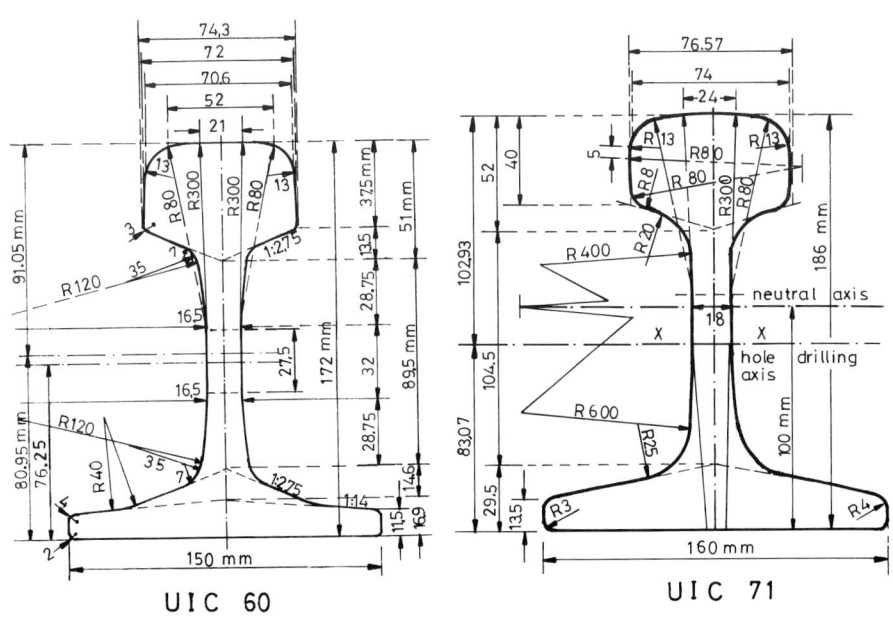

Fig. 5.5. Rail profiles UIC 60 and UIC 71

5.3. Rail steel grade, mechanical strength and chemical composition

5.3.1. Mechanical strength

The increase in train speed and per-axle load has led to the improvement of the grade of steel used for rails. The ultimate tensile strength was 50 kg/mm² in 1882, while today it is $70 \div 120$ kg/mm². A large increase in rail steel mechanical strength, however, may cause brittle failure and as a result a further strength increase is not desirable.

The rail steel grade may be distinguished in two categories:
- Normal steel grade, with $0.40 \div 0.50\%$ carbon content and an ultimate tensile strength of $70 \div 90$ kg/mm².
- Hard steel grade, used mainly on curves, with an ultimate tensile strength of $90 \div 120$ kg/mm².

5.3.2. Chemical composition

Concerning their chemical composition, rails present a great variety, (119):

5.3.2.1. Carbon

Increased carbon content increases hardness and wear resistance but at the expense of ductility and hardness. Normal quality steels contain $0.4 \div 0.6\%$ carbon.

5.3.2.2. Manganese

All commercial steels contain a small quantity of manganese ($0.3 \div 0.8\%$). Manganese in excess of this, increases hardenability. Increasing manganese and reducing carbon can achieve an equivalent tensile strength but with improved ductility. $11 \div 14\%$ manganese produces steels that are characterized by a great wear resistance (see section 5.8 below).

5.3.2.3. Chromium

Chromium increases hardness and wear resistance. Steels containing $2.0 \div 2.5\%$ chromium and $0.3 \div 1.5\%$ carbon are very hard and have a high tensile strength combined with a considerable degree of toughness and wear resistance.

5.3.2.4. Chromium-Manganese

The deleterious effect of increased carbon on the fatigue strength of steel can be moderated by using more manganese and chromium.

5.3.2.5. *Equivalent carbon percentage*

The related effects of carbon, manganese and chromium can be added together to produce an equivalent carbon percentage, given by the formula

$$\text{Equivalent carbon} = C\% + \frac{Mn\%}{3} + \frac{Cr\%}{3} \qquad (5.1)$$

It is found that an increase of 0.1% equivalent carbon raises tensile strength by 7 kg/mm².

The related effect of the three elements on wear resistance is not clear, but it has been recorded that an increase of 0.1% in equivalent carbon reduced vertical head wear (see section 5.8 below) by $4.5 \div 7.5\%$.

5.3.3. *Hard steel grades*

Hard steel grades have a minimum tensile strength of 90 kg/mm² and are widely used where lateral wear, plastic deformation and rail defects (see section 5.7) would shorten the life of normal quality rails.

There are three grades of hard steel rails, (119):

Grade	Carbon	Manganese
A	$0.6 \div 0.75\%$	$0.8 \div 1.3\%$
B	$0.5 \div 0.65\%$	$1.3 \div 1.7\%$
C	$0.45 \div 0.6\%$	$1.7 \div 2.1\%$

Grade C is difficult and slow to weld and for this main reason it has been universally rejected by the European railway administrations.

5.4. Stress analysis of rail

The total stresses developed on the rail are the sum of:
- Hertz stresses (at the wheel-rail contact)
- stresses resulting from rail bending on the ballast
- stresses resulting from bending of the rail head on the web
- stresses resulting from thermal effects
- plastic stresses, remaining on the rail after the removal of external loads.

With the exception of the last category, all others will be calculated on the assumption of an elastic behaviour. As discussed in chapter 4, section 4.3.3.2, both theory and experiments show that in most of the cases, rail has an elastic behaviour.

5.4.1. Stresses at wheel-rail contact

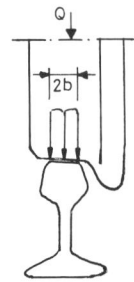

Fig. 5.6. Wheel-rail contact stresses according to Eisenmann

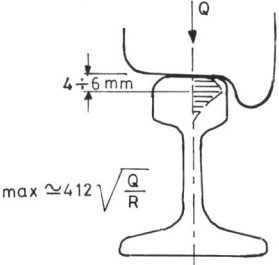

Fig. 5.7. Shear stresses at the wheel-rail contact

The problem of stresses developed at wheel-rail contact was examined by Dang Van, (110), in accordance with Hertz's assumption that the contact surface between two curved elastic bodies (wheel-rail head, see Fig. 5.6) is elliptical in shape and the stress distribution along the contact surface is semi-elliptical. Measurements have shown, however, that for wheel diameters between 60 and 120 cm (covering the majority of cases), the following two-dimensional simplified approximation gives satisfactory results (Eisenmann's theory).

Assuming that all radii of curvature (with the exception of the wheel radius R (in mm)) are infinite and that the wheel load Q (in Nt) is uniformly distributed, the mean Hertz stress is derived according to the Eisenmann analysis from the relation, (111)

$$\sigma_\mu (Nt/mm^2) = \sqrt{\frac{\pi E}{64(1-\nu^2)} \cdot \frac{Q}{Rb}} \quad (5.2)$$

Substituting the usual values of $E=2.1 \cdot 10^6$ kp/cm², $\nu=0.3$, $b=6$ mm, the following equation is derived:

$$\sigma_\mu = 1374 \sqrt{\frac{Q}{R}} \quad (5.2a)$$

The Eisenmann's simplified approximation, (111), gives the shear stress distribution of Fig. 5.7 and the maximum value:

$$\tau_{max} \simeq 412 \sqrt{\frac{Q}{R}} \quad (5.3)$$

The maximum stress according to Hertz occurs at a depth of $4 \div 6$ mm from the rolling surface.

5.4.2. Bending stresses of the rail on the ballast

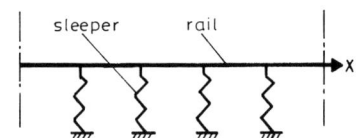

Fig. 5.8. Simulation of rail for the calculation of bending stresses

The rail is simulated by a continuous beam on elastic supports.
The general equation of mechanics:

$$EI \frac{d^4u}{dx^4} + ku = 0 \qquad (5.4)$$

gives the following analytical solution for the specific simulation:

$$\sigma_b = \frac{Q(h_r - z)}{4\gamma_r I_r} e^{-\gamma_r x} (\sin\gamma_r x - \cos\gamma_r x) \qquad (5.5)$$

where:
Q = the load per wheel
I_r = the rail moment of inertia in the vertical direction
h_r = the distance between rolling line and neutral axis of the rail
k = the track index = r/z, where r is the theoretically uniformly distributed load of the wheel on the rail and z is the rail vertical settlement

$$\gamma_r = \sqrt[4]{\frac{k}{4E I_r}}$$

5.4.3. Bending stresses of the rail head on the rail web

The rail head is simulated as a beam laying on an elastic sub-base.
The resulting stresses conform to the analytical expression, (121):

$$\sigma_h = \frac{Q(h_c - z)}{4\gamma_c I_c} e^{-\gamma_c x} (\sin\gamma_c x - \cos\gamma_c x) \qquad (5.6)$$

where:
I_c = the moment of inertia of the rail head, $\gamma_c = \sqrt[4]{\frac{I_c}{4}}$
h_c = the distance between rolling line and neutral axis of the rail head

5.4.4. Stresses caused by thermal effects

These are derived from the well-known equation:

$$\sigma_{th} = \alpha \cdot E \cdot \Delta\theta \qquad (5.7)$$

where α : the rail thermal expansion coefficient
Δθ : the temperature difference

5.4.5. Plastic stresses

Up to the present, no satisfactory elastoplastic analysis for the calculation of plastic stresses within rails has been conducted. A fundamental reason for this has been the difficulty in simulating limit conditions between the rail and the sleeper and in determining numerical values for the parameters of the elastoplastic calculation.

Measurements have yielded a plastic stress distribution as illustrated in Figs. 5.9 and 5.10.

Laboratory tests performed by the Japanese Railways on 50T-type rails (weighing 53 kg/m), (125), have shown a stress distribution as illustrated in Fig. 5.11. Similar results were obtained by the German Railways for rail type S49 (weighing 53 kg/m), (116).

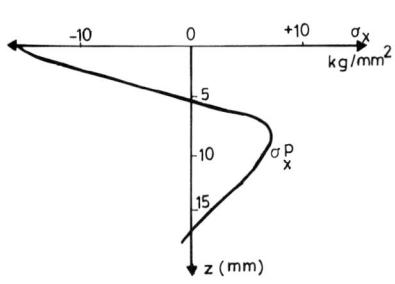

Fig. 5.9. Longitudinal plastic stresses σ_x^p at the plane of symmetry of the rail

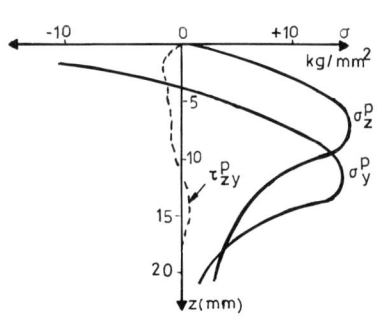

Fig. 5.10. Transverse plastic stresses σ_y^p, σ_z^p, τ_{zy}^p at the plane of symmetry of the rail

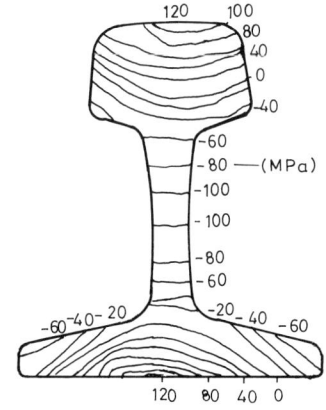

Fig. 5.11. Plastic stresses in rail type 50T (weighing 53 kg/m), (125)

5.5. Analysis of the mechanical behaviour of the rail by the finite-element and the photoelasticity methods

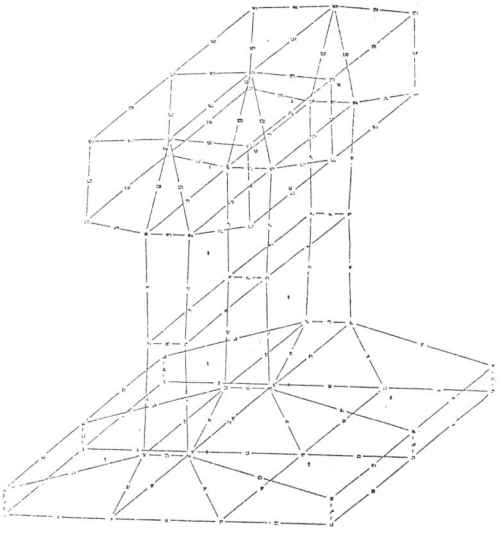

Fig. 5.12. Rail analysis by the finite-element method

The mechanical behaviour of rail may also be approximated by the finite-element method (Fig. 5.12), (91). In such a simulation, however, it is very hard to study the limit conditions at the rail-sleeper contact accurately. Therefore, it is customary to include both the rail and the sleeper in the finite-element analysis.

Unilateral contact and inequality mechanics theories, (122), can be implemented for the accurate study of the rail-sleeper contact, but research under progress, (121a), did not conclude significant results.

Finally, in order to investigate stress distribution of the rail, photoelastic methods may also be used. Figure 5.13 illustrates the shear stress distribution based on the photoelasticity method, (1).

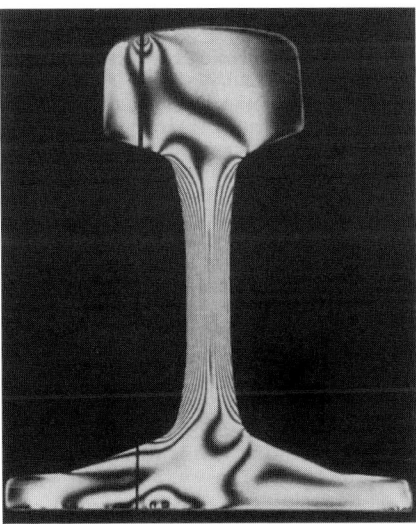

Fig. 5.13. Rail analysis using the photoelasticity method, (1)

5.6. Rail fatigue

5.6.1. Fatigue curve and Miner's rule

Fatigue can be defined as the gradual decrease of mechanical strength in a material under the influence of repeated loading as long as the developed stress exceeds a minimum value σ_0, known as the fatigue limit. For stresses below the fatigue limit ($\sigma < \sigma_0$) fatigue phenomenon does not occur.

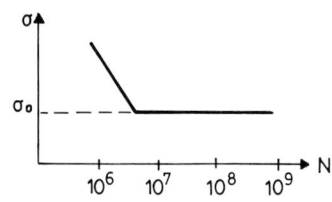

Fig. 5.14. **Fatigue curve (Wöhler)**

If stresses exceed the fatigue limit ($\sigma > \sigma_0$), then the mechanical strength gradually decreases, leading to fracture of the material for strength values lower than the fracture stress corresponding to the first loading cycle.

Theoretical and experimental research of the fatigue phenomenon mainly center on two topics:
  a. Determination of the fatigue curve (also known as the Wöhler curve). The fatigue phenomenon occurs for stresses $\sigma > \sigma_0$ (Fig. 5.14).
  b. For a stress history within the fatigue area, determination of the strength reserves of the material. Let σ_1 be a loading history, $\sigma_1 > \sigma_0$, for which the lifetime, at the end of which material failure will occur, is N_1 loading cycles. The material is subjected to n_1 cycles of loading, and $n_1 < N_1$. Let σ_2 be a second loading history, $\sigma_2 > \sigma_0$, which, in the absence of the σ_1 loading, would have made the lifetime N_2 loading cycles. Unknown is the number n_2 loading cycles, which will lead to material failure. The answer is given by the Miner rule with the approximative relation

$$\frac{n_1}{N_1} + \frac{n_2}{N_2} \simeq 1 \qquad (5.8)$$

In the case of more loading histories, the Miner rule is generalized as follows:

$$\frac{n_1}{N_1} + \frac{n_2}{N_2} + \frac{n_3}{N_3} + \ldots \frac{n_n}{N_n} \simeq 1 \qquad (5.9)$$

The origin of the fatigue phenomenon in metals involves internal discontinuities between crystals, which are present from the beginning. If the developed stresses are sufficiently small, these internal discontinuities do not propagate and thus the state of equilibrium is maintained. However, when stresses exceed the fatigue limit, then internal discontinuities propagate, expand, merge, and may cause failure of the material from fatigue without any visible macroscopical deformation.

5.6.2. Dang Van's rail fatigue criterion

The rail fatigue phenomenon has been extensively researched, both at the experimental, (114), and at the theoretical, (110), level, so as to investigate the conditions leading to the commencement of instability due to an inherent internal discontinuity. On the basis of the finding that internal discontinuities tend to propagate towards grains with crystallographic planes less well oriented to resist external loads, and taking into consideration a series of laboratory test results, Dang Van formulated a criterion, named after him, according to which rail fatigue develops in two phases, (110):

1. A first hardening phase, during which stresses develop under the influence of cyclic plastic strains and tend to a limiting equilibrium state. Assuming isotropic hardening of crystal steel, it can be shown that local stresses $\sigma_{ij}(t)$ are connected to macroscopic stresses $\Sigma_{ij}(t)$ (those resulting from the continuum theory) by the relation:

$$\sigma_{ij}(t) = \Sigma_{ij}(t) - a_{ij} T_o \quad (5.10)$$

where α_{ij} : the grain orientation tensor

with m : the sliding direction
    n : perpendicular to the sliding plane
    T_o : the mean shear, defined for the n-cycle as

$$T_o^n = \frac{1}{2} (T_{max}^n + T_{min}^n) \quad (5.11)$$

2. A second phase during which propagation of internal discontinuities starts in grains that are already in a plastic state, while surrounding grains are in an elastic region. Since the number of molecules remains constant, the creation of internal voids results in an increase in volume, a fact justifying the investigation of the role of the spherical (or hydrostatic) tensor $(\sigma_{KK}/3)$* in the study of the rail fatigue phenomenon,

$$\sigma_{ij} = \frac{\sigma_{KK}}{3} \delta_{ij} + s_{ij}$$

s_{ij} being the deviator tensor and δ_{ij} Kronecker's delta

---

* It is worth recalling that a repeating subscript means the sum at all possible values of the subscript (Einstein's assumption). Thus, (108):

$$\frac{\sigma_{KK}}{3} = \frac{\sigma_{11}}{3} + \frac{\sigma_{22}}{3} + \frac{\sigma_{33}}{3}$$

Macroscopic stresses $\Sigma_{ij}(t)$ result from the continuum theory, while experimental findings determine n, m, and therefore the tensor α_{ij}.

Local shear $\tau(t)$ in grains with the worst orientation will be:

$$\tau(t) = T(t) - T_o \qquad (5.12)$$

where T : the macroscopic shear
T_o : the mean shear

On the basis of the foregoing equations, the $\tau - \dfrac{\sigma_{\kappa\kappa}}{3}$ diagram, also known as the Dang Van diagram, can be plotted (Fig. 5.15).

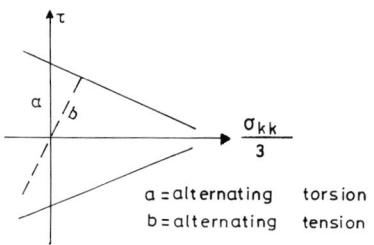

Fig. 5.15. **Dang Van's criterion for rail fatigue**

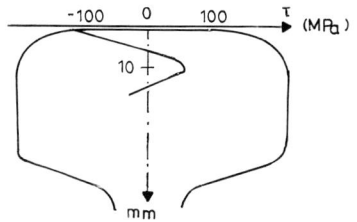

Fig. 5.16. **Shear stress variation within a rail**

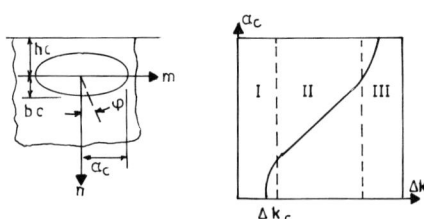

Fig. 5.17. **Evolution of an internal discontinuity**

Analysis of the rail fatigue phenomenon has shown that, (110):
* Maximum shear stress develops $10 \div 15$ mm below the rolling surface (Fig. 5.16). It should be noted that this conclusion has been confirmed by a series of laboratory tests, (114).
* Maximum stresses occur in planes inclined 30° to the vertical.
* An increase in wheel diameter causes an increase of the internal discontinuities.
* Internal discontinuities causing fatigue are proportional to per-axle load Q raised to a power between 3 and 4 and closer to 4.

5.6.3. Evolution of an internal discontinuity

The evolution of an internal discontinuity, elliptical in form, with major axis $2a_c$, is a function of the stress intensity Λk exerted on the discontinuity perimeter. For values $\Lambda k < \Lambda k_c$, discontinuity dimensions remain unaffected by external loading. The region II (Fig. 5.17) is where discontinuity shows a large increase, estimated, (112), by the relation:

The Rail

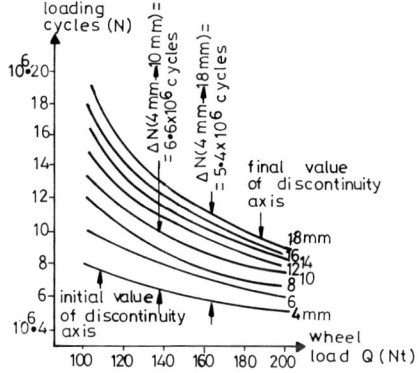

Fig. 5.18. Evolution of a discontinuity as a function of wheel load Q and the number N of loading cycles

$$\frac{d\alpha_c}{dN} = c \cdot \Lambda k^{n_c} \qquad (5.13)$$

where c and n_c are constants resulting from laboratory tests. Finite-element analysis enables the calculation of the number of loading cycles making an initial discontinuity reach a particular value as a function of wheel load, (116) (Fig. 5.18).

A more empirical relation giving the evolution of internal discontinuities as a function of traffic load T of the line has been derived from research conducted by the ORE, (114):

$$Y = Y_o \cdot 2^{T\,5} \qquad (5.14)$$

where
Y_o : initial value of the discontinuity
T : line traffic load in million tons per year

A rail runs a serious risk of fracture when the expanding and merging internal discontinuities cover more than 55% of the surface of the head of the rail, (131).

5.7. Rail defects

Internal discontinuities which may give rise to rail fatigue are called rail defects. Rail alterations of a mechanical nature occurring under the influence of passing trains are also considered defects. *Rail defects* should be clearly distinguished from *track defects*, the latter being defined as the deviations of actual from theoretical values of the track's geometric characteristics. Track defects, (172), are exclusively the consequence of train traffic and are of a macroscopic and geometric nature, usually reversible by track maintenance (see chapter 11, section 11.3). On the contrary, rail defects are due to initial manufacturing imperfections of the rail, are of a mechanical and microscopic nature and in most cases non-reversible.

Rail defects have been studied and classified by the International Railways Union (UIC), (126). The principal rail defects, which are the cause of the most serious risks of fatigue failure, include the following:

Railway Engineering

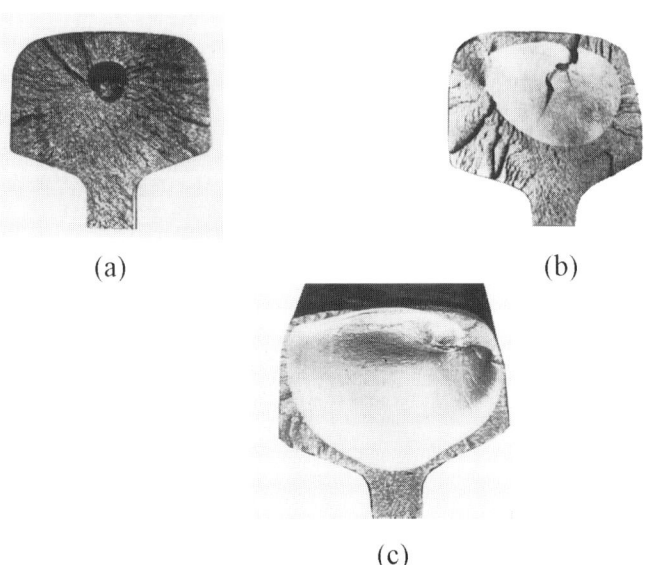

(a)   (b)

(c)

Fig. 5.19. Tache ovale

5.7.1. Tache ovale (Rail defect 211 according to the UIC, Fig. 5.19). This corresponds to an initial internal oval discontinuity, caused by thermal effects during rail manufacture. It expands to reach the rail surface and causes immediate failure of the rail. It may be at the origin of very serious problems and even reach epidemic proportions in rails of the same manufacture. It is detected either by visual inspection or with the aid of ultrasonic equipment. Most research work on rail fatigue centers on this defect, (126).

5.7.2. Horizontal cracking (Rail defect UIC 212, Fig. 5.20), referring to horizontal cracks of the rolling surface of the rail. It originates at the manufacturing stage (initial internal discontinuities) and may cause peeling of the rolling surface. It is detected either visually or by ultrasonic equipment.

5.7.3. Short-pitch corrugations (Rail defect UIC 2201, Fig. 5.21). Their cause is train traffic and they consist of corrugations with wavelength $\lambda = 3 \div 8$ cm. They can provoke many adverse effects: high frequency oscillation of the track, including resonance and leading to higher rail stresses, concrete sleeper fatigue with cracking in the rail seat area, loosening of fastenings, accelerated wear of pads insulators and clips, premature failure of ballast and the subgrade, an increase by $5 \div 15$ dB (A) in noise level. This defect is detected either visually or by rail defect recording equipment. It is repaired by passage of special equipment which grinds and smooths the rail, (126), (114).

5.7.4. Long-pitch corrugations (Rail defect UIC 2202). They have wavelengths $\lambda = 8 \div 30$ cm and occur mainly on the low rails of curves having a radius of

The Rail

Fig. 5.20. Horizontal cracking

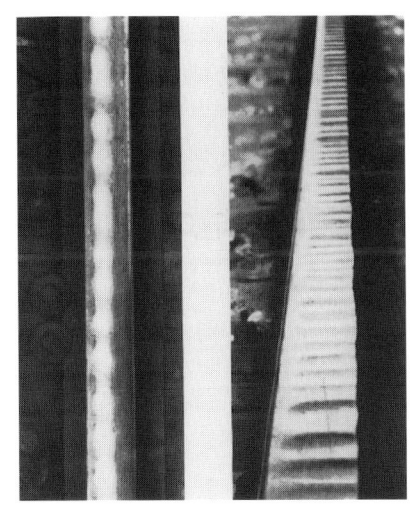

Fig. 5.21. Short-pitch corrugations

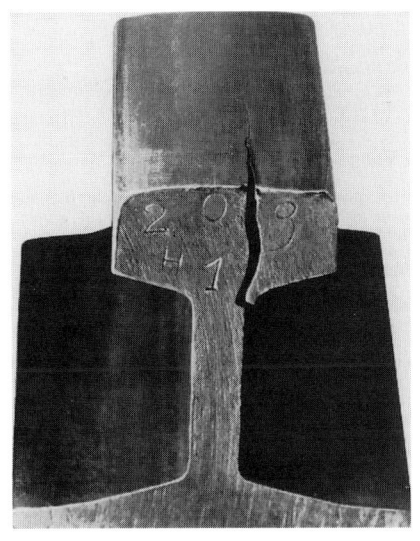

Fig. 5.22. Longitudinal vertical cracking

Fig. 5.23. Lateral wear

600 m and smaller. This form of wear is most common on suburban and underground railways carrying a large volume of identical traffic. Detection and repair are as per the short-pitch corrugations.

5.7.5. Longitudinal vertical cracking (Rail defect UIC 113, Fig. 5.22), causing vertical cracks, which may expand and split the rail head in two. This is again a rail manufacture defect. It is detected by ultrasonic equipment and the affected rail should be immediately replaced.

5.7.6. Lateral wear (Rail defect UIC 2203, Fig. 5.23). This is lateral rail head wear caused by the snaking course of the wheels of the rolling stock. Lateral wear becomes dangerous beyond a certain point because it affects the track gauge. The various railway networks specify the permissible value of lateral rail head wear (see section 5.8).

5.7.7. Rolling surface disintegration (Rail defect UIC 221, Fig.5.24), corresponding to a gradual disintegration of the rolling surface of the rail. The causes of this defect can be traced to the rail manufacturing process. The defect is detected during maintenance inspections and affected rails are replaced at scheduled maintenance sessions.

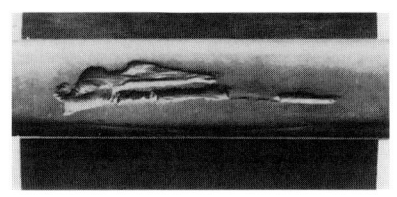

Fig. 5.24. Rolling surface disintegration

5.7.8. Shelling of the running surface (Rail defect UIC 2221). Irregular deformation of the running surface is observed prior to the formation of shells several millimeters deep in the metal. The cross-section of these shells is extremely variable. Shelling is not an isolated defect. It always occurs over a wide area. Detection is done either visually or by ultrasonic testing.

Fig. 5.25. Gauge-corner shelling

5.7.9. Gauge-corner shelling (Rail defect UIC 2222, Fig. 5.25). The rails first show long dark spots randomly spaced out over the gauge corner of the rail head. These spots are early signs of underlying metal disintegration which, after a period of evolution, are characterized by the formation of lips on the side face, of cracks and lastly of shelling in the gauge corner which can sometimes be quite extensive. This form of shelling usually occurs along the outside rails in curves lubricated to avoid lateral wear.

5.8. Permissible rail wear

5.8.1. Vertical wear

The maximum permissible vertical wear of the rail is a function of the maximum train speed and of the line traffic load. Tables 5.2 and 5.3 give the maximum permissible wear values of the rail head according to the British and the German Railways, (119).

It should be noted that rail wear by locomotive wheels is about 6 times heavier than that from the wheels of tracted rolling stock.

Table. 5.2.
Maximum permissible vertical wear of rail (159 mm high) according to the British Railways, (119)

Maximum speed (km/h)	Maximum permissible vertical wear of the rail head (mm)
> 160	9
120 - 160	12
80 - 120	15
< 80	18

Table 5.3.
Maximum permissible vertical wear of rail (154 mm high) according to the German Railways, (119)

Kind of line	Maximum permissible vertical wear (mm)
Lines with an annual load exceeding 19 million tons or daily load exceeding 25,000 tons or speeds exceeding 140 km/h or more than 120 trains per day	12
Lines with annual load exceeding 7.5 million tons or with daily load between 20,000 and 25,000 tons	20
Lines with annual load exceeding 1.75 million tons	26

5.8.2. Lateral wear

The maximum permissible lateral wear according to the British Railways is defined in accordance to a reference point located 3 mm above the lowest point of the rail head and at a 26° angle to the rail axis (Fig. 5.26a).

The German Railways define the maximum lateral wear (Fig. 5.26b) on the MN line, where M is the gravity center of the rail head. On main tracks and for UIC 60 profile, lateral wear should not exceed 18 mm. The sum of the vertical and the lateral wear of the rail head, however, should not exceed 25 mm, (119).

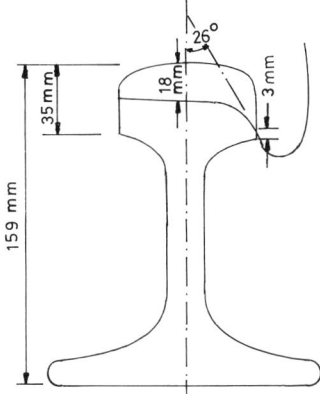 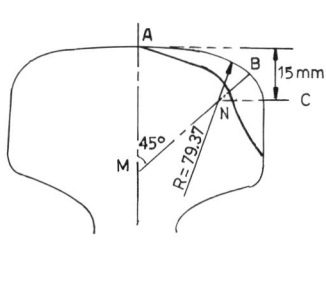

Fig. 5.26a. Maximum permissible lateral wear of rail according to the British Railways

Fig. 5.26b. Maximum permissible lateral wear of rail according to the German Railways

5.9. Optimum lifetime of a rail

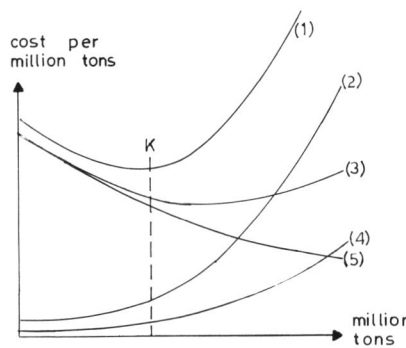

(1) Total cost
(2) Cost from derailments
(3) Total cost excluding derailments
(4) Cost of repairs-wear
(5) Cost of rails

Fig. 5.27. Optimum rail lifetime

Determining the optimum lifetime of a rail is not a purely technical problem but rather a techno-economic one. Beyond a service period of the rail, the total cost increases sharply (Fig. 5.27). It is therefore advisable to replace the rail before all technical strength margins are exhausted. Optimal rail lifetime is determined by point K (as per Fig. 5.27), corresponding to a minimum in total cost, (124). However, a rail removed from a principal line can be used for a certain period on secondary lines.

For a UIC 60 rail profile the German Railways assume a service life of 40 years for principal lines and 100 years for their secondary lines (in practice these times are 39 and 82 years respectively), (8).

The Rail

For an S 54 rail profile (weight 54.5 kg/m of length), the German Railways predict a service life as follows:
- First service life (principal lines):   24 years
- Second service life (secondary lines):   32 years
- Third service life (other lines):   38 years

Total service life: 94 years

For a number of reasons, however, (damage, exceptional wear, corrosion), the potential second service life is reduced by 25% and the potential third service life by 75%, thus leading to a total service life of 57 years, (8).

French Railways achieve an average service life of 55-60 years and the British Railways around 45 years. Once again, the best of the serviceable rail recovered is cropped, welded and re-used for lower category lines.

5.10. Fishplates

Until about 30-40 years ago, tracks were laid in all networks (and are still laid in many networks) by leaving gaps between consecutive rails, and then connecting the rails with fishplates (Fig. 5.28). The basic purpose of the gaps was to absorb length variations due to temperature fluctuations.

The fishplate connection technique was detrimental to rail transportation in several ways:
- it significantly reduced passenger comfort,
- it caused considerable wheel and rail fatigue and wear,
- it greatly increased maintenance expenses, on the one hand due to the necessary inspections to ensure proper condition of all fishplate parts, and on the other because of the height irregularities arising in the fishplate region.

In standard-gauge lines, fishplates are usually installed every 36 or 54 m, after prior welding of the rails in groups of two or three. A characteristic of fishplate-connected tracks is the limited contraction-expansion capability of rails depending on temperature fluctuations.

Fig. 5.28. Fishplates connecting rails

Railway Engineering

Every rail type has a corresponding fishplate type, as well as a particular form of bolt. Figure 5.29 illustrates the fishplate arrangement for UIC 54 rails, and Fig. 5.30 shows the bolts for fishplate type UIC 54.

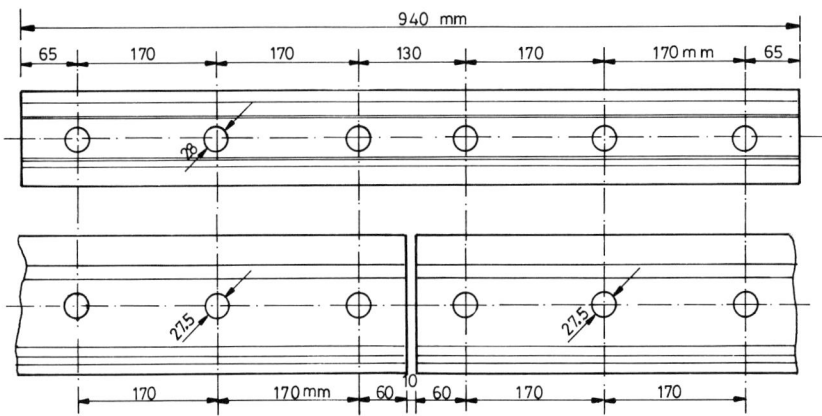

Fig. 5.29. Fishplate perforation for UIC 54 rail

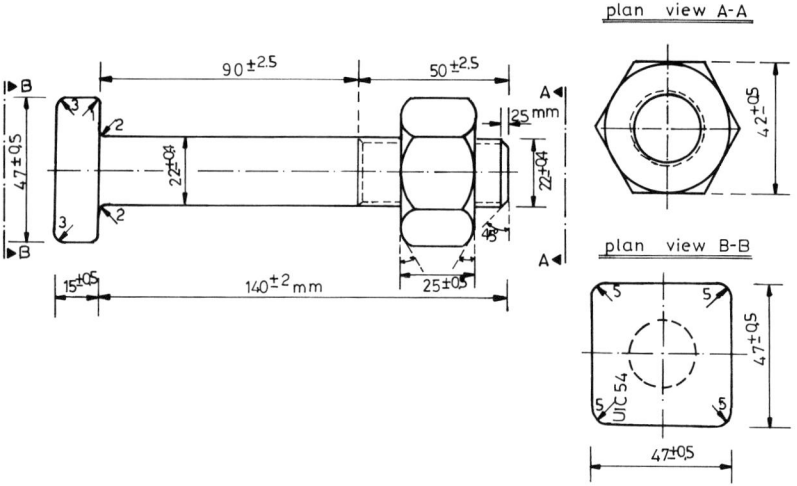

Fig. 5.30. Fishplate bolt type UIC 54

5.11. The continuous-welded rail

5.11.1. The continuous-welding technique

From the time when railways were first introduced, efforts were made to increase the length of rails, the ultimate goal being a continuous track. A continuous-welded rail (c.w.r.) is the result of welding together discrete pieces of rail stock as obtained from the production plant in various lengths, commonly 18, 24, 30 or 36 m for standard-gauge tracks. The usual maximum length of production of rails is 36 m (U.K., France, Italy, etc.), but may attain greater values in some countries (60 m in Germany, up to 108 m in Austria, etc.). In contrast to fishplate-joined rails, c.w.r.'s are characterized by a rail region where no temperature-induced length change occurs. Continuous welding does away with fishplates, with all the obvious beneficiary consequences this entails.

Although it is a technically simple concept, continuous welding took a long time to be adopted in railway technology. This delay was due to the following reasons, (9), (120):

* As previously mentioned, a characteristic of the continuous-welded rail is the absence of length variation. This is a result of the friction forces between sleeper and ballast as well as between rail and sleeper. These forces, however, cannot be taken for granted unless the rail-sleeper connection is permanent and reliable. This has been enabled over the last 40 years by elastic fastenings (see chapter 6, section 6.9.2.2), which ensure that thermal stresses are uniformly distributed and constant in time.
* The fatigue behaviour of welds which undergo repeated stresses by the passage of trains was not adequately known. Recent research has shed light on this aspect and there are now no reservations concerning this matter.
* Finally, the risk of buckling was also considered due to the great length of the c.w.r. Research on the mechanical resistance of the ballast, which opposes buckling, combined with a track weight increase, have addressed the problem in a satisfactory manner.

In the case of tramway lines, which are fully restrained by being embedded in the road, the problem of a built up of longitudinal forces does not occur and longer rails could be implemented.

5.11.2. Mechanical behaviour of the continuous-welded rail

5.11.2.1. Assumptions

Non-linear constitutive laws, numerical models and the knowledge of fatigue mechanisms, are the scientific areas the development of which in recent years has contributed to a more accurate approach of the mechanical behaviour of

Railway Engineering

the c.w.r., at the price, however, of increasingly complex calculations. Accordingly, a simplified theory is presented which, although leading to conclusions deviating from the accurate theory, is easily grasped, gives a faithful qualitative representation of actual phenomena in addition to giving safety-oriented results, (120).

It is assumed that the behaviour of all materials is *elastic* and that ballast resistance is uniform and constant.

5.11.2.2. Calculation procedure

The continuous-welded rail is simulated by a bar with length L and a cross-sectional area S (Fig. 5.31). Under the influence of a temperature difference $\Delta\theta$ the bar length changes will be:

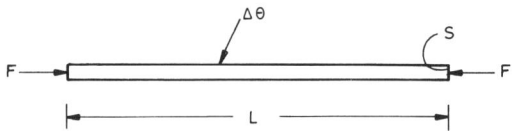

Fig. 5.31. Simplified simulation for studying the behaviour of the continuous-welded rail

$$\Delta\ell^{\Delta\theta} = \alpha \cdot L \cdot \Delta\theta \tag{5.15}$$

where α is the thermal expansion coefficient of steel.

The ballast resists the change in length caused by $\Delta\theta$ by a force F. According to the simplified assumptions used, the change in length due to F will be:

$$\Delta\ell^F = \frac{F \cdot L}{E \cdot S} \quad \text{(Hooke's law)} \tag{5.16}$$

The overall stress situation will result from the superposition of the two forces (of opposite direction) previously mentioned, and therefore:

$$\Delta\ell^{tot} = \alpha \cdot L \cdot \Delta\theta - \frac{F \cdot L}{E \cdot S} \tag{5.17}$$

We need to find a force F such that the change in length $\Delta\ell^{tot}$ will be zero. From equation (5.17) it follows that:

$$F = \alpha \cdot E \cdot S \cdot \Delta\theta \tag{5.18}$$

a relation showing that the ballast force F is independent of rail length and proportional to the cross-sectional area. The latter proves that the forces developed depend on the rail type.

From the aforementioned equation it can be calculated that a force of 1.85 tons per degree centigrade is generated in the case of UIC 60 section and 1.6 tons for UIC 54 section.

5.11.2.3. Force distribution along continuous-welded rail

Forces generated along a c.w.r. by thermal effects are transmitted through the fastenings and sleepers to the ballast. Let r denote the ballast resistance with values ranging from 0.5 to 1.0 ton/metre of track. This resistance is obviously zero at the end of the track and, cumulatively increasing over a length ℓ_A (Fig. 5.32), it generates a resultant force equal to F. According to equation (5.18) therefore, it will be

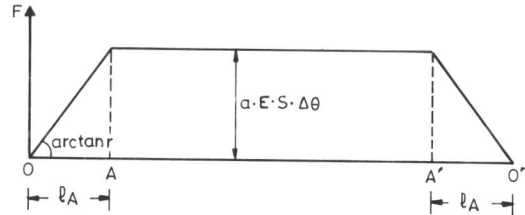

Fig. 5.32. Diagram of forces developed within c.w.r. (0=c.w.r. near end; 0'=c.w.r. far end)

$$r \cdot \ell_A = F = \alpha \cdot E \cdot S \cdot \Delta\theta \qquad (5.19)$$

from which

$$\ell_A = \frac{\alpha \cdot E \cdot S \cdot \Delta\theta}{r} \qquad (5.20)$$

The length ℓ_A corresponds to what in railway engineering is often referred to as the expansion zone. Beyond this length the resultant force, due to the ballast resistance, completely balances out the force developed by thermal effects along the c.w.r. Therefore, beyond the length ℓ_A, no length change takes place.

Considering an average value r = 0.75 t/m for the ballast resistance and the case of UIC 60 rail (S=76.86 cm²), we will have for $\Delta\theta$=35°C:

$$\ell_A = 85.3 \text{ m} \qquad (5.21)$$

which in certain cases may reach a maximum value in the order of 150 m.

Since the length of the c.w.r. cannot be smaller than $2\ell_A$ (because if it were, no point of the c.w.r. would remain immobile during changes in temperature), it follows from the foregoing that the minimum length of a c.w.r. is 300 m.

Railway Engineering

5.11.2.4. Length changes in the expansion zone

The c.w.r. undergoes a change in length due to temperature variations only in the expansion zone ℓ_A, beyond which every c.w.r. point remains immobile. The displacement of the point O (see Fig. 5.32), caused by the superposition of stresses generated by temperature variation and ballast resistance is calculated as follows:

i. Due to a temperature change $\Delta\theta$, a length change $\Delta\ell_{\ell_A}^{\Delta\theta}$ will occur:

$$\Delta\ell_{\ell_A}^{\Delta\theta} = \alpha \cdot \ell_A \cdot \Delta\theta = \alpha \cdot \Delta\theta \frac{\alpha \cdot E \cdot S \cdot \Delta\theta}{r} = \frac{\alpha^2 \cdot E \cdot S \cdot \Delta\theta^2}{r} \qquad (5.22)$$

ii. Due to ballast resistance, the value of which is zero at point O and r_A at point A, a length change $\Delta\ell_{\ell_A}^{\Delta\theta}$ will occur. Assuming a linear distribution, there will be a resultant force equal to $r\ell_A/2$, producing a displacement

$$\Delta\ell_{\ell_A}^r = \frac{r\ell_A}{2} \frac{\ell_A}{ES} = \frac{r}{2ES}\ell_A^2 = \frac{r}{2ES}\left(\frac{\alpha \cdot E \cdot S \cdot \Delta\theta}{r}\right)^2 = \frac{\alpha^2 \cdot E \cdot S \cdot \Delta\theta^2}{2r} \qquad (5.23)$$

Combining equations (5.22) and (5.23), we obtain:

$$\Delta\ell_{\ell_A}^{tot} = \Delta\ell_{\ell_A}^{\Delta\theta} + \Delta\ell_{\ell_A}^r = \frac{\alpha^2 \cdot E \cdot S \cdot \Delta\theta^2}{r} - \frac{\alpha^2 \cdot E \cdot S \cdot \Delta\theta^2}{2r} \Rightarrow$$

$$\Delta\ell_{\ell_A}^{tot} = \frac{\alpha^2 \cdot E \cdot S \cdot \Delta\theta^2}{2r} = k\Delta\theta^2 \qquad (5.24)$$

where $k = \alpha^2 ES/2r$ is constant for a ballast with a given mechanical strength.

5.11.2.5. Rail welding

The welding of rail steels is generally achieved by flash-butt or electrical resistance welding (usually in depots) followed by in situ welding into track using one of the thermit welding processes.

5.11.2.5.1. Flash-butt welding

The flash-butt welding process is a method of joining metals in which the heat generated, necessary to forge the joint, is created by the resistance of the rails being welded to the passage of an electric current. Unlike the thermit welding process, no additional chemicals or metals are required to make the weld. In flash welding the parent metal is consumed during the welding cycle, thus creating the necessary heat along the rail ends in order to accomplish the merging action and consolidate the joint. A total length of approximately 25 ÷ 35 mm, depending upon the rail section, is consumed per weld.

Flash welding can be carried out at:
- fixed site depots
- mobile depots
- in track

5.11.2.5.2. Thermit welding

The thermit welding process is based on the reduction of heavy metals from their oxides with the aid of aluminium. This reaction is strongly exothermic, since a very large quantity of heat is generated, and is brought about by the strong affinity which aluminium exhibits towards oxygen.

There are two main iron oxides, FeO and Fe_2O_3 which react with Al according to the following equations:

$$Fe_2O_3 + 2Al \rightarrow 2Fe + Al_2O_3 + 181.5 \text{ Kcals} \tag{5.25}$$

$$3FeO + 2Al \rightarrow 3Fe + Al_2O_3 + 186 \text{ Kcals} \tag{5.26}$$

In the thermit steel welding, it is important to suitably control the Al content:
- an excess of Al between 0.2 ÷ 0.65% is desirable, but above 0.7% it may cause the steel to solidify
- a deficiency of Al in the melting process can cause heat shortness due to the creation of steel constituents such as C and Mn, reacting with the Fe oxides because of the lack of Al.

Many European administrations use the German thermit process called SKV - a welding process with short preheating, (8).

In every welding procedure, the appropriate control of welds is critical for the longevity of the c.w.r.

5.11.2.6. Destressing of a continuous-welded rail

It is desirable that c.w.r. welding and laying be carried out at a temperature ranging between the upper and lower extremes, so as to minimize c.w.r. stress.

Regardless of the c.w.r. laying temperature, however, the reduction of thermal stresses is sought. This is achieved by destressing the c.w.r. and creating free expansion (or contraction) conditions. Destressing is done after an elapse of time from the c.w.r. laying, depending upon the traffic load necessary to stabilize the track. This load is usually 100,000 tons in the case of timber sleepers and 20,000 tons for concrete sleepers.

Destressing should be done gradually along 800 ÷ 1,000 m and exceptionally on 1,200 m track lengths. The following procedure can be implemented, (120):

Railway Engineering

i. If the c.w.r. is longer than 1,200 m, destressing is done in sections. The rails are cut at the end of each section and the ends are diverted to enable free rail movement.
ii. Fastenings are loosened.
iii. Rails are placed on rollers (Φ 20 every $10 \div 20$ sleepers), so as to reduce friction as much as possible.
iv. Further reduction of the restraining friction is achieved by lateral blows along the rail by wooden or plastic sledge hammers.
v. If at the time of destressing, rail temperature is less than the mean temperature of the country, the rail is heated (by propane heaters) to reach the optimum mean temperature, desirable to minimize stresses at extreme temperatures. Obviously, if the rail temperature exceeds the mean temperature, no additional heating is required.
vi. The rollers are removed and the fastenings tightened.
vii. Destressing should take place on both rails. On each track section, destressing works should be performed during traffic-free intervals.

However, good maintenance inspection is necessary on all c.w.r. to ensure that are avoided both buckles in hot weather and rail breakages through brittle fructure in cold weather.

5.11.3. Expansion devices

The length variation at the end of a c.w.r. was calculated in section 5.11.2.4. In order to ensure that these length variations will not be accompanied by excessive stresses at certain sensitive points along the track (e.g. ends of steel bridges, station entrances-exits, etc.), expansion devices (e.d.) are installed at these points.

Figure 5.33 illustrates the details of an e.d. for UIC 54 rail, while Fig. 5.34 shows the overall track arrangement in the vicinity of an e.d. There is a great variety of such device types among the railway networks.

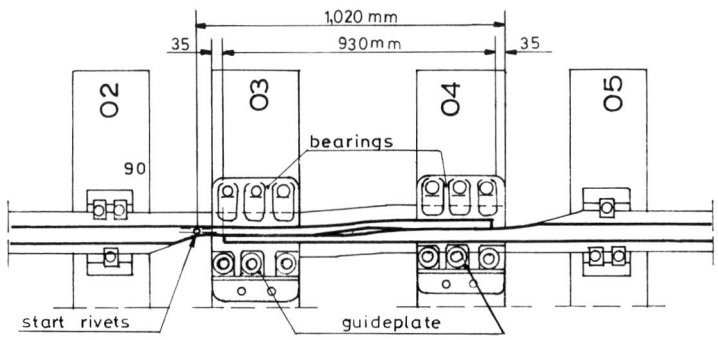

Fig. 5.33. Expansion device for UIC 54 rail (all dimensions in mm)

The Rail

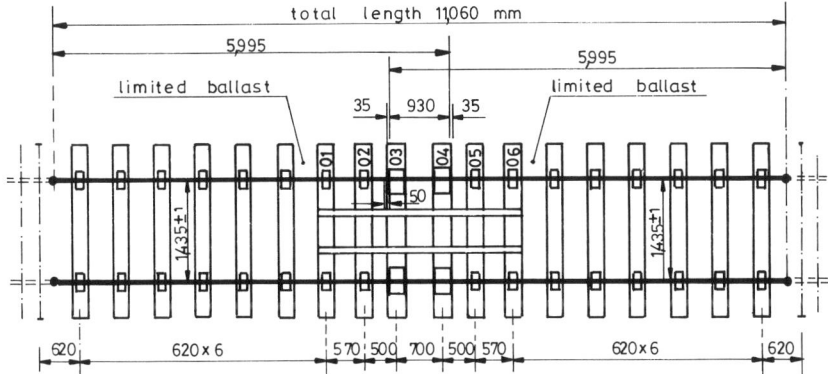

Fig. 5.34. Track arrangement in the region of an expansion device (all dimensions in mm)

Expansion devices should not be used in the following cases:
- on transition curves between straight and curved track,
- on curves with small radius of curvature (less than 800 m),
- within 120 m from level crossings in the case of rails weighing 50 kg/m or more and within 100 m for rails of less than 50 kg/m,
- on large bridges without ballast:
  • if the bridge is more than 20 m long, expansion devices are required at each end
  • if the bridge is less than 20 m long, c.w.r. may be laid with no e.d.

5.11.4. Advantages of the continuous-welded rail

Although the cost of installing c.w.r. is higher than with fishplated track, an adequate return on capital for the initial investment is provided by reduced maintenance cost of the permanent way, improved track stability, higher running speeds, lower power consumption and improved passenger comfort. In particular:
  • c.w.r. offer a much higher comfort level
  • the development of track defects is much slower with c.w.r.
  • track equipment fatigue is smaller
  • loads developed in wheels and on the rolling stock are generally much lower.

6 Sleepers - Fastenings

6.1. The various types of sleepers

Sleepers are the track components positioned between rails and ballast. The rails of the first railway lines were mounted on blocks placed directly on the ground. The need for better load distribution led to the addition of the sleepers and ballast.

The sleepers must ensure the following:
- appropriate load transfer and distribution from the rails to the ballast
- constant rail spacing, as specified by the track gauge
- mounting of the rails on the sleepers at an inclination of 1/20 or 1/40
- adequate mechanical strength both in the vertical and in the horizontal direction.

Along electrified lines, sleepers should moreover ensure (either individually or with added accessories) the electrical insulation of each rail from the other.

The first material used for sleepers was wood. Its scarcity and sensitivity led to the introduction of steel sleepers around 1880, which were widely used for a long time. Advances in concrete technology led to the use of concrete sleepers after 1950, distinguished in two categories:
- twin-block reinforced-concrete sleepers
- monoblock prestressed-concrete sleepers.

Presently, more than three billion sleepers are installed in various rail networks worldwide, of which about 500 million are concrete sleepers. $2 \div 5\%$ of the total number of sleepers are replaced each year, (134).

Sleepers presently installed along new lines or overhauled old ones are mostly of concrete. However, timber sleepers are also used in several instances. The use of steel sleepers is diminishing and they are usually replaced by concrete or timber sleepers at track renewals or construction of new tracks.

Choice of the most appropriate sleeper type should be made for each line by feasibility analysis which includes an evaluation and assessment of the following factors (while reducing preference due to prejudice):

Sleepers - Fastenings

- sleeper's construction or purchasing cost
- purchasing cost of fastenings and other indispensable sleeper accessories
- sleeper lifetime
- maintenance cost
- probable residual value of sleeper at the end of its lifetime.

6.2. Steel sleepers

6.2.1. Form and properties

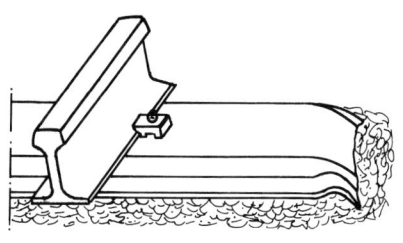

Fig. 6.1. Steel sleeper

The steel sleeper is an industrial product of simple construction. It consists of a rolling type profile in the form of ∩. Its ends are forged to provide anchoring in the ballast, so as to ensure transverse track stability (Fig. 6.1).

The rail is mounted on to the steel sleeper by rail spikes (crampons) fixed by rail spike bolts in holes drilled onto the sleeper top. Elastic fastenings may also be used.

6.2.2. Manufacturing, size and weight

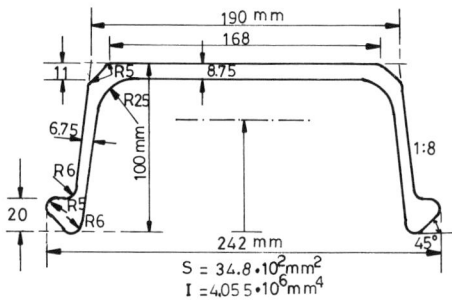

Fig. 6.2. Cross-section of steel sleepers with the help of computer aided design (dimensions in mm)

Steel sleepers are made from low carbon steel of an ultimate tensile strength of $40 \div 50$ kg/mm². Generally sophisticated steels have not been used and therefore the yield strength is near 50% of ultimate strength. The chemical composition is usually 0.15% C, 0.45% Mn, $0.01 \div 0.35$% Si, $0 \div 0.35$% Cu.

Computer programs have helped in recent years to optimize the cross-section of steel sleeper (Fig. 6.2). If we compare the moment of inertia of this new cross-section with older ones, we observe that with the same quantity of material the stiffness of the sleeper has almost doubled.

Fig. 6.3a gives the geometrical characteristics of a steel sleeper (used in tracks for low speeds, V<140 km/h), weighing 70÷80 kg. At the area of rail joints, where a greater steel resistance is needed, a twin-type steel sleeper (Fig. 6.3b) can be used, weighing 130÷140 kg.

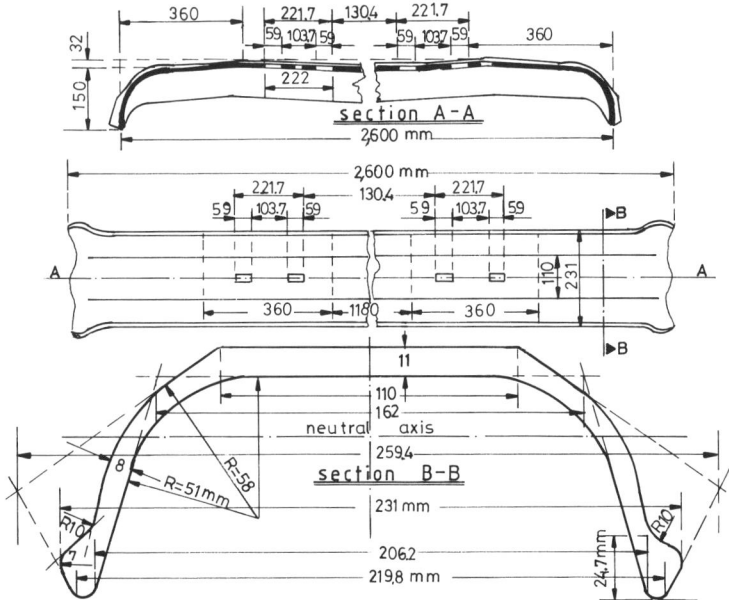

Fig. 6.3a. Steel sleeper, simple type

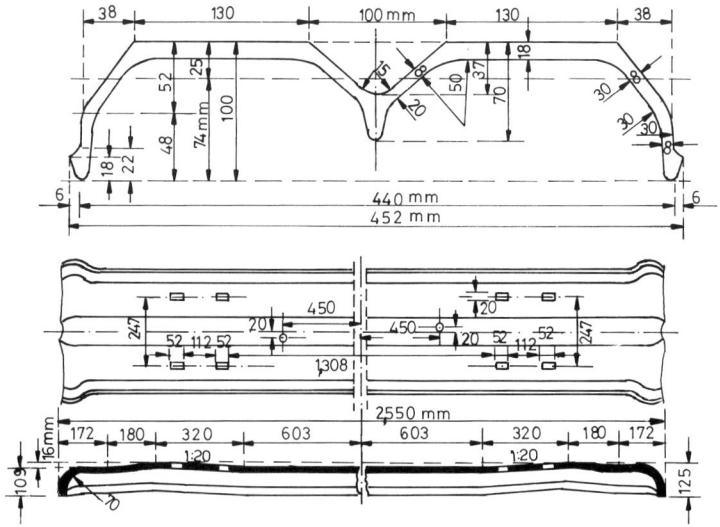

Fig. 6.3b. Steel sleeper, twin-type

6.2.3. Advantages and disadvantages

Steel sleepers are rather easily manufactured, installed and maintained. They keep the track gauge adequately constant for a long time and their transverse resistance is considerable. Their lifetime is relatively long and after replacement they have still a certain value as scrap iron.

However, steel sleepers have many disadvantages. Their form makes longitudinal and transverse track positioning difficult. Steel sleepers are noisy, they require special insulating devices for signalling, and their maintenance is difficult. Furthermore, steel sleepers are sensitive to chemical attack and particularly vulnerable in lines close to industrial and coastal areas. All the above disadvantages have led to the economic devaluation and to the gradual withdrawal of steel sleepers, particularly in Europe.

6.2.4. Lifetime

Steel sleeper lifetime ranges between 30 and 60 years with a mean value of 50 years, (137a).

6.3. Timber sleepers

6.3.1. Form and properties

Timber sleepers distribute loads better than other sleeper types. They are accordingly indicated for tracks laid on fair or poor quality subgrade, where concrete sleepers would require a comparatively thicker ballast layer. Timber sleepers had been used extensively for 100 years or more. Because of their higher cost and shorter lifetime, their use in Europe is presently limited to instances where concrete sleepers are not indicated. However, they are still extensively used in North America.

The kinds of wood presently used for timber sleepers include beech and oak from European trees, and azobé from tropical ones. Pine tree timber has also been used in the past. Timber sleepers in use by the various railways today are mostly of azobé tropical timber, which is stronger and more durable. In underground tunnels, Australian jarrah hardwood sleepers have been used extensively.

To extend its lifetime, a timber sleeper is impregnated with special fluids, commonly coal tar. Impregnation is carried out by a specific process developed for the various kinds of timber.

In order to prevent timber sleeper splintering or slippage on the ballast, it is necessary to contain the wood fiber within the ballast. This is achieved by suitable configuration of the sleeper ends, which are either braced by a steel strap surrounding the sleeper end, or have special metal plates driven into the vertical section of the sleeper ends.

Railway Engineering

Timber sleepers are particularly sensitive and their strength decreases with time as a result of:
- deterioration of their mechanical characteristics
- influences of a chemical nature
- influences of a biological nature.

6.3.2. Dimensions

The geometrical characteristics of timber sleepers are specified in UIC Technical Directive 863. Timber sleepers in *standard-gauge* lines have typical dimensions as shown in Fig. 6.4, (55).

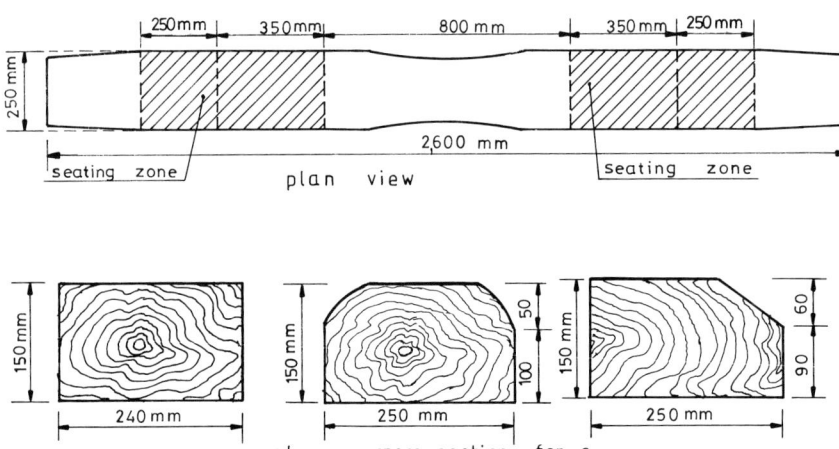

Fig. 6.4. Geometrical characteristics (dimensions in mm) of timber sleepers in standard-gauge lines

The following tolerances are allowed in the above dimensions:
Length: +40 mm, −30 mm
Width: −10 mm
Height: −5 mm

In *metric-gauge* lines, timber sleepers have the dimensions shown in Fig. 6.5.

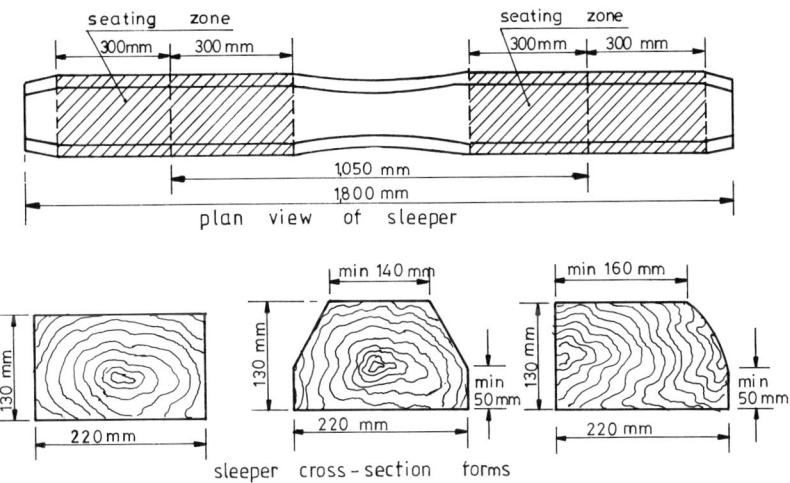

Fig. 6.5. Timber sleeper characteristics (dimensions in mm) in metric-gauge lines

The following tolerances are applicable to the above dimensions:
Length: +30 mm, −30 mm
Width: 0
Height: +/− 5 mm

6.3.3. Advantages and disadvantages

The principal advantage of timber sleepers is flexibility and the resulting better load distribution. Timber sleepers are accordingly mainly indicated in the case of poor subgrade. Timber sleepers moreover provide good insulation and do away with special devices for signalling and electric traction. Finally, compared to concrete sleepers, timber sleepers are shorter in height.

The disadvantages of timber sleepers include their relatively short lifetime, their higher cost and their low transverse resistance (a result of their low weight) precluding high train speeds.

6.3.4. Lifetime

The lifetime of timber sleepers depends on the wood type used and is:
- 25 years for oak timber (impregnated)
- 30 years for beech timber (impregnated)

Railway Engineering

- 40 years for azobé tropical timber (non-impregnated)
- 45 years for azobé tropical timber (impregnated)
- 50 years for jarrah or similar hardwood used in tunnels.

6.3.5. Timber sleeper deformability

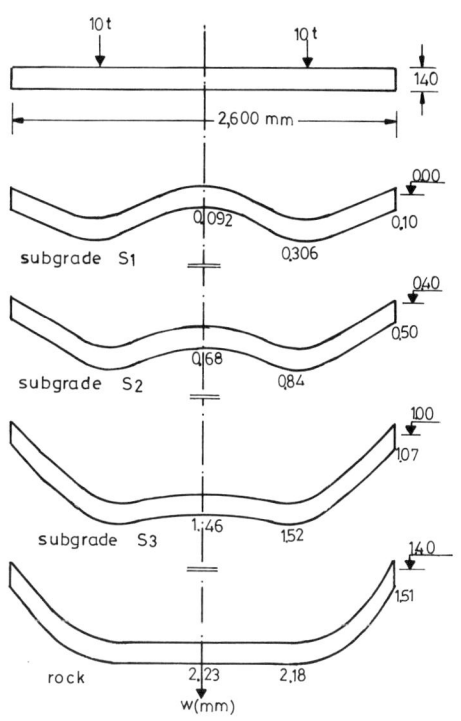

Fig. 6.6. **Deformability of timber sleeper for various qualities of the subgrade, (142)**

Finite-element analysis (see chapter 4, section 4.3) provides a precise and detailed determination of timber sleeper deformability for various soil subgrade qualities (see chapter 3, section 3.4) and various track support thicknesses e (e = ballast + sub-ballast). Figure 6.6 illustrates the deformability of the timber sleeper on poor-quality subgrade (S_1), on medium-quality subgrade (S_2), on good-quality subgrade (S_3) and on rock (R), (142). It is observed that the poorer the subgrade soil quality is, the more uniform is timber sleeper settlement.

6.4. Concrete sleepers

6.4.1. Inherent weaknesses of concrete sleepers

Monoblock reinforced-concrete sleepers when first introduced, after 1920, presented the following serious intrinsic weaknesses:
- A propensity for brittle fracture under the influence of dynamic train loads, and extensive cracking, leading to failure.
- Very little fatigue resistance resulting in high tensile stresses in the central

part of the sleeper, which, if the tensile strength exceeded, led to slippage of the reinforcing bars.

To overcome these two weaknesses it would require:
- Laying the rails so that they did not have direct contact with the sleepers, by interposing an absorbing material to blunt load impact. Such material includes rubber pads, which in turn necessitate the use of elastic fastenings.
- Using inexpensive reinforcing bars with the same lifetime as concrete.

6.4.2. The two types of concrete sleepers

In tandem with the reinforced-concrete and the prestressed-concrete technologies, two concrete sleeper types were developed:
- The twin-block reinforced-concrete sleeper, consisting of two trapezoidal reinforced-concrete sections joined by a connecting bar (Figs 6.7, 6.8)
- The monoblock prestressed-concrete sleeper, which can be pretensioned or post-tensioned (Figs 6.10, 6.11).

Since the load distribution beneath the sleeper (see section 6.7) has shown that in the central section the developing stresses are very small, it follows that less material can be safely used in this part of the sleeper. As a result, in the central part of the twin-block sleeper the concrete was replaced by a connecting bar (which principally serves to maintain the track gauge), while in the prestressed sleeper (where the above solution cannot be applied) the cross-section of the central part of the sleeper was reduced.

The twin-block sleeper was developed in France and the principal users are: Algeria, Belgium, Brazil, Denmark, Greece, Mexico, Netherlands, Portugal, Spain, Tunisia, etc.

The monoblock pretensioned sleeper was developed in the United Kingdom and the principal users are: Australia, Canada, Hungary, Iraq, Japan, Norway, Poland, South Africa, Sweden, USA, Russia.

The monoblock post-tensioned sleeper was developed in Germany and the principal users are: Austria, Finland, India, Italy, Mexico, Turkey.

Of all new concrete sleeper installations, twin-block account for 20% and monoblock 80%. The total number of concrete sleepers made each year throughout the world is about 20 million, (134).

The use of concrete sleepers on curved tracks is a controversial subject. In their metric gauge tracks, South African Railways are not using concrete sleepers in curves below 300 m radius. On the other hand, Canadian Railways, who experience very extreme temperatures ($-40°C$ to $+30°C$), are installing concrete sleepers in all curves of radius less than 870 m including many less than 200 m, with continuous-welded rail and no gauge widening. To some extent the different approach may arise from shortcomings in certain fastenings, (56a).

Railway Engineering

6.5. The twin-block reinforced-concrete sleeper

6.5.1. Geometry and mechanical strength

Figure 6.7 shows the geometrical characteristics of the twin-block reinforced-concrete sleeper U 31 of the French Railways which weighs 180 kg. The connecting bar has a Y or L shaped section. This sleeper can be wholly produced in each country, except perhaps the bar which is imported, and is implemented on medium-load lines of UIC 3, 4 and 5 groups (see chapter 2, section 2.5.2). It allows speeds of up to 200 km/h for axle loads of 19÷21 t and up to 220 km/h for axle loads of 17 t, (137).

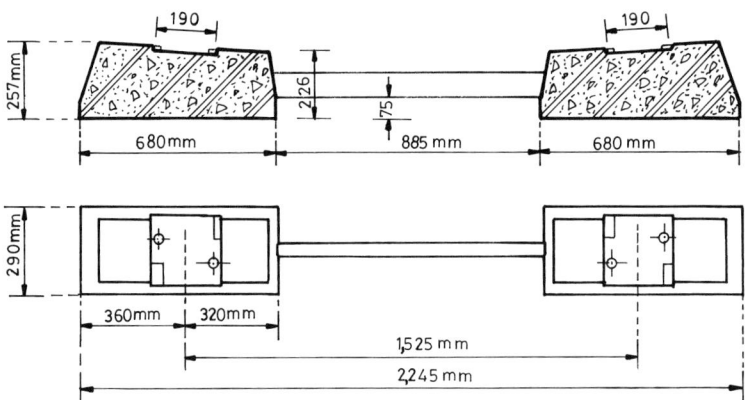

Fig. 6.7. Twin-block reinforced-concrete sleeper U31 of the French Railways used in groups UIC 3, 4 and 5 lines (dimensions in mm), (137)

Twin-block sleepers require ballast thickness and strength greater than that required by timber sleepers. Whenever this requirement was met, twin-block sleepers gave very satisfactory results.

Particular care should be exercized when the twin-block sleeper is seated on poor-quality subgrade. In this case the ballast thickness should be further increased.

In more heavily loaded lines (UIC 1, 2 groups) and speeds > 200 km/h, a different, larger type of twin-block sleeper, is used. Figure 6.8 presents the geometrical characteristics of such a sleeper of the French Railways, known as U41. This sleeper weighs 260 kg and has been used at the TGV tracks which are run at a speed of 300 km/h, (137).

Because of the flexible tie bar, twin block sleepers require extra maintenance in service, so as to ensure that the blocks do not tilt differentially and loose gauge.

Sleepers - Fastenings

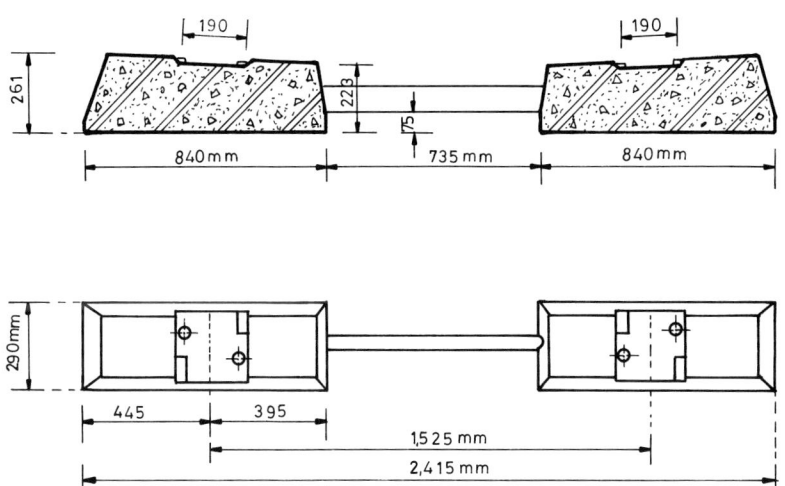

Fig. 6.8. Type U41 twin-block sleeper of the French Railways (for groups UIC 1 and 2 and speeds up to 300 km/h) (dimensions in mm), (137)

6.5.2. Advantages and disadvantages

Due to its great weight, the twin-block sleeper provides very satisfactory transverse track resistance and allows high train speeds. It keeps track gauge within satisfactory tolerances and has a long lifetime. It can be produced in each country and generally is less expensive than timber sleepers.

Twin-block sleeper behaviour is less satisfactory when the ballast does not have the suitable thickness and mechanical characteristics. Load distribution and flexibility are less satisfactory with twin-block than with timber sleepers. In addition, twin-block sleepers require elastic fastenings and because of their great weight, handling is difficult. The twin-block sleeper (in contrast to the timber sleeper) requires special accessories, so as to ensure the necessary insulation for signalling and electric traction operation. Special attention should be given to the behaviour of the tie bar. If the tie bar is not appropriately placed, it may produce a maintenance hazard to staff working on the track, which is overcome in the monoblock sleeper.

6.5.3. Lifetime

The twin-block sleeper has a lifetime of around 50 years.

6.5.4. Deformability of the twin-block sleeper

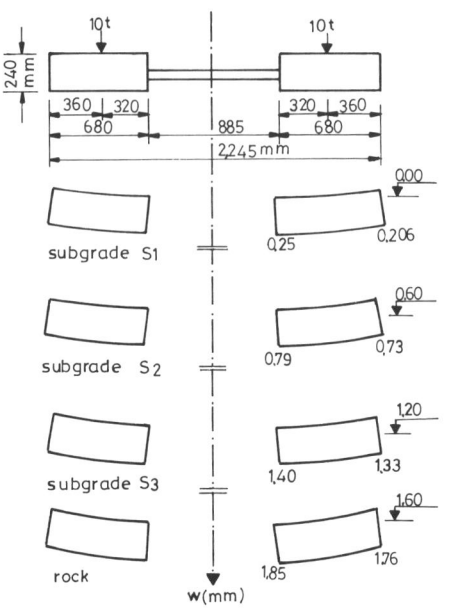

Fig. 6.9. Deformability of the twin-block sleeper for various subgrade qualities, (142)

Figure 6.9 illustrates the deformability of the U 31 twin-block sleeper for various qualities of the subgrade (S_1, S_2, S_3, R) and track support thicknesses, (142). It is observed that deformability is much lower than that of timber sleepers. Accordingly, in the case of a poor-quality subgrade, the use of twin-block sleepers should be accompanied by an increase of ballast thickness, which should have adequate mechanical strength.

6.6. The prestressed-concrete monoblock sleeper

6.6.1. Geometry and mechanical strength

The geometrical characteristics of the monoblock prestressed-concrete sleeper

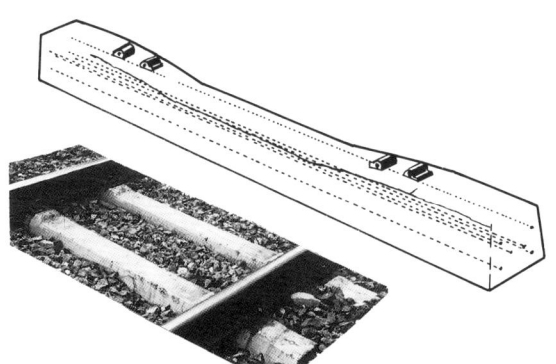

Fig. 6.10. Monoblock prestressed-concrete sleeper

(Fig. 6.10) are similar to those of the timber sleeper and its mechanical strength is similar to that of the twin-block sleeper. The monoblock sleeper:

- withstands alternating stresses better, since the stress on the concrete is always compressive

- offers a reduced sleeper height at the central part, since the steel bars do not have to be located, as in reinforced-concrete, as far away from the neutral axis as possible
- allows reduction of the steel used, in comparison to the twin-block sleeper
- generally is lighter compared to the twin-block sleeper; a fact, however, which also reduces transverse resistance.

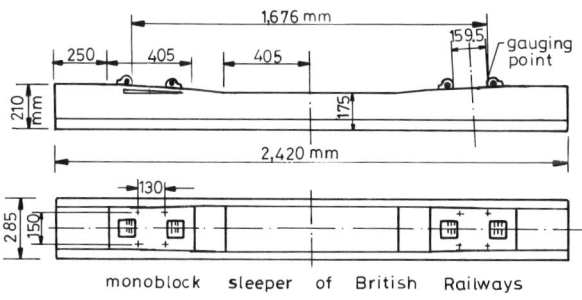

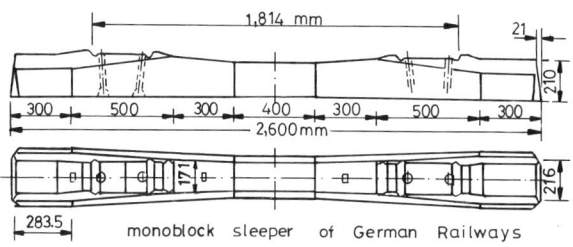

Fig. 6.11. Monoblock sleepers of the British and German Railways

Monoblock sleepers come in a large variety of geometrical features. All, however, are characterized by a reduction of the cross-section at the central region. Figure 6.11 illustrates the geometrical characteristics of the monoblock sleepers of the British Railways (with initial prestressing force 38.9 t and residual prestressing force 32.1 t) and of the German Railways (with a weight of 280 kg, initial prestressing force 32.5 t and residual prestressing force 27.0 t) Table 6.1 gives the geometrical characteristics of the monoblock sleepers in use by principal rail networks all over the world and Table 6.2 presents the mechanical characteristics of monoblock sleepers on various rail networks, (134). A critical element in monoblock sleeper design is the ratio λ of the maximum moment M_{cr}, which the sleeper can withstand, to the maximum moment M_{max} developing in the sleeper. The factor in question takes values between 1.3 and 1.8.

6.6.2. Advantages and disadvantages

Monoblock sleepers have a behaviour similar to that of the twin-blocks. They maintain the track gauge in a satisfactory manner and have a long lifetime. They require elastic fastenings and special accessories for signalling.

However, monoblock sleepers distribute loads better than twin-blocks, but not as well as timber sleepers. Their transverse resistance is lower than that of twin-blocks, but higher compared to timber sleepers; monoblock sleepers provide also a good surface for maintenance inspection staff.

Table 6.1.
Geometrical characteristics of monoblock prestressed-concrete sleepers used in various rail networks, (134)

Country	Track gauge (mm)	Sleeper length (mm)	Sleeper dimensions (mm) Rail position			Sleeper dimensions (mm) Sleeper center		
			H	W_B	W_T	H	W_B	W_T
Australia	1,435	2,500	212	250	200	165	250	200
Canada	1,435	2,542	203	264	216	159	264	226
China	1,435	2,500	203	280	170	165	250	161
Germany	1,435	2,600	214	300	170	175	220	150
India	1,673	2,750	210	250	n.a.	180	220	n.a.
Italy	1,435	2,300	172	284	222	150	240	190
Japan	1,435	2,400	220	310	190	195	236	180
Russia	1,520	2,700	193	274	177	135	245	182
South Africa	1,065	2,057	221	245	140	197	203	140
Sweden	1,435	2,500	220	294	164	185	230	150
U.K.	1,432	2,515	203	264	216	165	264	230
U.S.A.	1,435	2,591	241	279	241	178	279	250

Table 6.2.
Mechanical characteristics of the various types of monoblock prestressed-concrete sleepers, (134)

Country	Sleeper spacing (mm)	Rail type	Maximum train speed (km/h)	Minimum radius of curvature (m)	Maximum load per axle (tn)	Length ℓ_{ex}* (m)	Maximum permiss. moment M_{max} (tm)	Permiss. stress in concrete (kg·cm)	Maximum moment developed M_{cr} (tm)	Coefficient $\lambda \cdot \frac{M_{cr}}{M_{max}}$
Australia	550-600	53 60 kg/m	160	200	24.5	0.53	1.62	23	2.38	1.5
Canada	610	132RE 136RE	130	194	29.2	0.55	2.01	33	3.06	1.5
China	550	50 kg/m	120	350	24.5	0.53	1.62	26	1.34	0.8
Germany	600-650	S54 UIC 60	250	100	22.1	0.58	1.60	30	1.84	1.2
India	650	UIC 60	130	550	22.0	0.54	1.49	20	2.43	1.6
Italy	600	UIC 60	180	485	22.1	0.43	1.19	47	1.50	1.3
Japan	590	50.4 60.8 kg/m	210	1.200	16.4	0.48	0.96	n.a.	1.73	1.8
Russia	500-643	R50 R65 R70	200	350	26.5	0.59	1.95	20	1.35	0.7
South Africa	700	48 47 kg/m	160	150	22.1	0.50	1.38	28	1.12	0.8
Sweden	600, 650	SJ50	130	300	22.2	0.53	1.47	30	1.50	1.0
U.K.	650, 700	BS113A	200	400	24.5	0.54	1.65	45	2.50	1.5
U.S.A.	610	65 69 kg/m	200	610	32.1	0.58	2.33	50	4.24	1.8

* See Fig. 6.13 (opposite page)

6.6.3. Lifetime

Monoblock sleeper's lifetime is around 50 years.

6.6.4. Deformability of monoblock sleepers

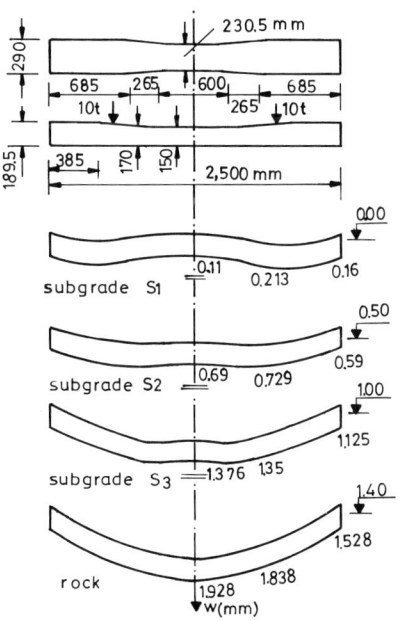

Fig. 6.12. Deformability of the monoblock sleeper on various subgrades

Figure 6.12 illustrates the deformability of one monoblock sleeper for various subgrade qualities (S_1, S_2, S_3, R) and track support thicknesses, (142). It is observed that the monoblock sleeper has a deformability similar to that of the timber sleeper but less flexibility. Monoblock sleepers should therefore be laid on ballast of suitable thickness and mechanical strength.

6.7. Stresses developing under the sleepers

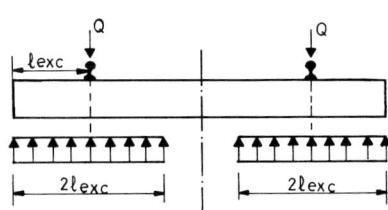

Fig. 6.13. Simplified sleeper model

The bending moments developed in sleepers may be studied by the simplified simulation of Fig. 6.13 where:
- the sleeper is simulated as a beam protruding at both ends,
- wheel load is assumed to be applied at a point,
- ballast reaction is considered uniform over a length $2\ell_{exc}$ below each rail.

Railway Engineering

However, the last assumption is not accurate. Analysis of the effects occurring at the sleeper-ballast interface is especially complex; it belongs to the unilateral contact problems of mechanics and at present no satisfactory approach has been found, (121a), (122).

On-site stress measurements under the sleeper have yielded the distribution of Fig. 6.14, with a maximum stress σ_1, given by the relation, (138)

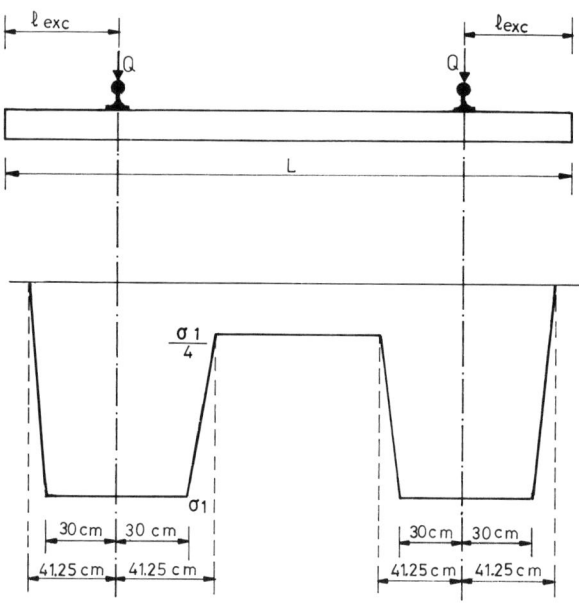

$$\sigma_1 = \frac{P}{\alpha \left(\frac{L}{2} + \frac{3\ell_{exc}}{2} \right)} \quad (6.1)$$

where
α : sleeper width
L : sleeper length
ℓ_{exc} : distance between sleeper end - wheel load application point
P : per-axle load
$P = 2Q$

Fig. 6.14. Stress distribution under sleeper, (138)

6.8. Track on concrete slab

6.8.1. The two forms of slab track

As previously mentioned in chapter 2, section 2.3, in addition to flexible ballasted track (seated on sleepers and ballast), slab track (inflexible seating) is also possible. Two forms of slab track can be distinguished:
* track seating on concrete slab by interposing sleepers (Photo 6.1)
* direct track seating on concrete slab (Photo 6.2).

Sleepers - Fastenings

Photo 6.1. Slab track by interposing sleepers

Photo 6.2. Slab track with no intervening sleepers

6.8.2. Problems created on a slab track

The concrete slab used for inflexible seating may be:
- of reinforced-concrete, in which case cracks were often found
- of prestressed-concrete, which offers a better mechanical behaviour and load distribution.

Once installed, it is costly to make vertical adjustments to a non-ballasted track to overcome settlements and therefore its use is confined to locations where a good quality subgrade can be provided.

Where lines are electrified, it is possible for stray currents to be induced into the reinforcement. To avoid electrolytic corrosion of the steel, the cement paste acting as the electrolyte, the bars are made electrically continuous by butt and spot welding the longitudinal and transverse bars together at regular intervals. The aim is to provide many leakage paths to the ground so as to minimize current density and hence prevent corrosion.

6.8.3. Advantages and disadvantages of slab track

Slab track has the advantage of ensuring very satisfactory transverse resistance and providing very high speeds. Maintenance cost is almost non-existent, but construction cost is higher, depending on the cost of the man-power in the specific country. For instance, it is reported that for the U.K. the cost of continuous slab track is 30% greater than the cost of excavating and replacing with conventional ballasted track. This additional cost could be recovered by savings in maintenance within a $5 \div 7$ years period, (144). Other countries, however, such

Railway Engineering

as France, reported greater differences in the construction cost of slab track compared to the ballasted track.

Slab track is particularly of interest in the case of tracks in tunnels, because it allows a reduction of the tunnel height (about 25 cm) and elimination of maintenance. In such cases measures to reduce track stiffness become necessary, and rubber pads of suitable thickness should be installed (see section 6.10 below). It is to note that the lack of resilience can cause vibration and noise problems coupled with early breakage of cast-in fixings.

Even in tunnels, however, a continuous slab track should be avoided in the case of medium or bad quality subgrades.

6.8.4. Geometrical characteristics of a slab track

Figure 6.15 illustrates the geometrical characteristics of a form of slab track in use by the British Railways, (144).

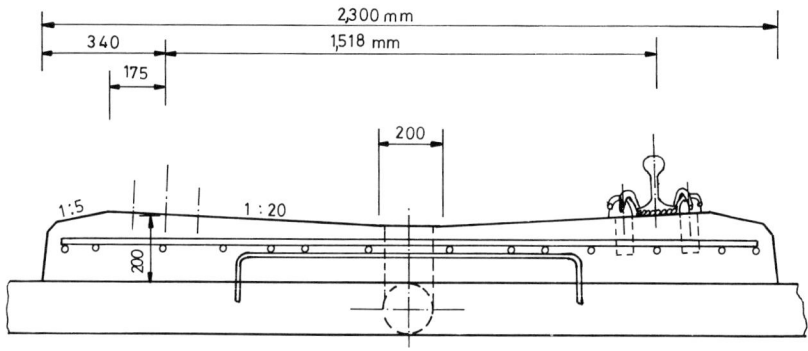

Fig. 6.15. Geometrical characteristics of a slab track (dimensions in mm), (144)

6.9. Fastenings

6.9.1. Functional characteristics

As fastenings are termed the set of parts and materials ensuring the rail-sleeper connection; they should provide the majority of the following properties:
- keep track gauge and transverse rail inclination on the sleeper constant
- transfer loads from the rail to the sleeper
- attenuate and dampen vibrations caused by train motion
- easy installation and maintenance
- electrical insulation
- resilience and adequate deflection

- avoidance of abrasion between components and of over-stressing
- adequate corrosion resistance
- reasonable cost and lifetime compatible to sleeper
- resistance to vandalism

6.9.2. Types of fastenings

Fastenings are distinguished in rigid and elastic fastenings.

6.9.2.1. Rigid fastenings

Rigid fastenings are used only with timber or steel sleepers. In rigid fastenings the rail is connected to the sleeper with bolts or nails. During train passage the rail compresses the sleeper and part of the strain is plastic (i.e. it does not disappear when the load disappears), resulting in the creation of a gap between nail head and rail. With continued train passage the gaps grow, causing a gradual slackening of the fastening, which affects safety and may be the origin of a derailment. In addition to plastic strain, high frequency vibration caused by rolling stock traffic also contributes to the widening of the gaps and the slackening of the fastening.

Rigid fastenings may be installed either without (Fig. 6.16) or with (Fig. 6.17) a seating plate, the latter being the preferable solution. With bull head rail rigid fastenings are provided by using cast iron chairs which are screwed or botted to sleepers, the rail being held by steel or timber keys or wedges.

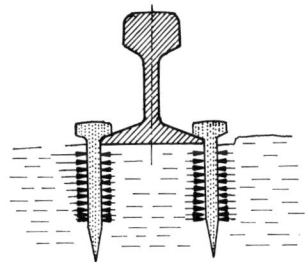
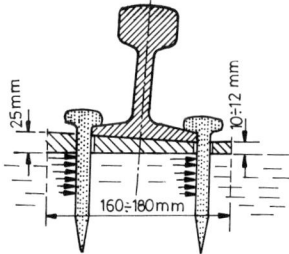

Fig. 6.16. Rigid fastening without seating plate

Fig. 6.17. Rigid fastening with seating plate

6.9.2.2. Elastic fastenings

Use of elastic fastenings is mandatory with concrete sleepers and optional with timber and steel sleepers. Two types of elastic fastenings may be distinguished:

Railway Engineering

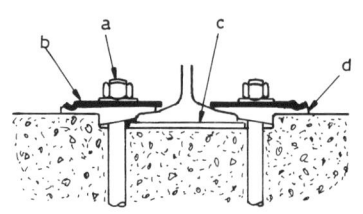

Fig. 6.18. Screw-type elastic fastening

- *Screw-type* elastic fastenings (Fig. 6.18). They have the advantage of great fastening strength and easy maintenance and replacement. They have the disadvantage that correct installation is affected by local conditions. Screw types are RN, Nabla, Vossloh and other fastenings. The common elements in all these designs are (Fig. 6.18):

- a threaded element (a) which is used to apply a force to a spring steel element, this threaded element being removable from the sleeper
- the spring steel element (b) which can be of bar or plate section
- a pad (c) between rail and sleeper to absorb vibration and impact, to provide a conformable layer between rail and sleeper and to provide electric insulation
- insulating elements (d) to isolate electrically the rail from any metallic path into the sleeper.

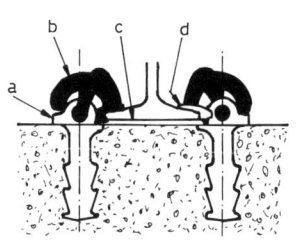

Fig. 6.19. Spring-type elastic fastening

- *Spring-type* elastic fastenings (Fig. 6.19). They are less adaptable than screw-type fastenings, but less affected by installation conditions, and any error is easily located visually. Pandrol, Lineloc, Hambo, etc. fastenings are of the spring-type. The common elements in spring-type fastenings (which should not require any subsequent maintenance) are (Fig. 6.19):

- some form of anchorage (a) fixed in the sleeper, generally at the time the sleeper is manufactured
- a spring steel element (b) to generate clamping forces on the rail foot
- a rail pad (c) to attenuate forces between rail and sleeper, to provide electrical insulation between rail and sleeper and to give conformable surfaces between rail foot and sleeper
- insulators or a layer of insulating materials (d), to provide electrical insulation between the rail and any metallic path, such as via (a) and (b), to the sleeper.

Sleepers - Fastenings

Photo 6.3. Nabla fastening   Photo 6.4. Pandrol fastening

6.9.2.3. Types of elastic fastenings

Elastic fastenings may also be installed with or without a seating plate. We will refer to all fastenings in use worldwide, shown numbered in Table 6.3 (next page). The following types of elastic fastenings can be identified:
- Fastenings with direct mounting without seating plate (numbered 6, 7, 8)
- Fastenings with indirect mounting without seating plate (numbered 9, 10, 11)
- Fastenings with direct mounting with seating plate (numbered 1, 2)
- Fastenings with indirect mounting with seating plate (numbered 3, 4, 5, 12, 13, 14).

Fastenings numbered 1 to 5 are older than fastenings numbered 6 and next.

6.9.2.4. Operating principles of elastic fastenings

During operation, elastic fastenings should ensure the following:
- The rail-sleeper fastening force should be sufficient to make the rail-sleeper slippage resistance much greater than the resistance to longitudinal motion of the sleeper on completely stabilized ballast.
- The fastening resonance frequency should be distinctly higher than the rail resonance frequency
- Fastenings should retain sufficient clamping force over the years
- Fastening tightness should be easily checked on the track without disassembling

Railway Engineering

- Fastenings should retain their elastic characteristics for a long time after installation
- Fastening efficiency (i.e. the ratio of the force applied on the rail base to the force transmitted by the fastening to the sleeper) should be as high as possible.

Table 6.3.
Various types of elastic fastenings used by rail networks, (1)

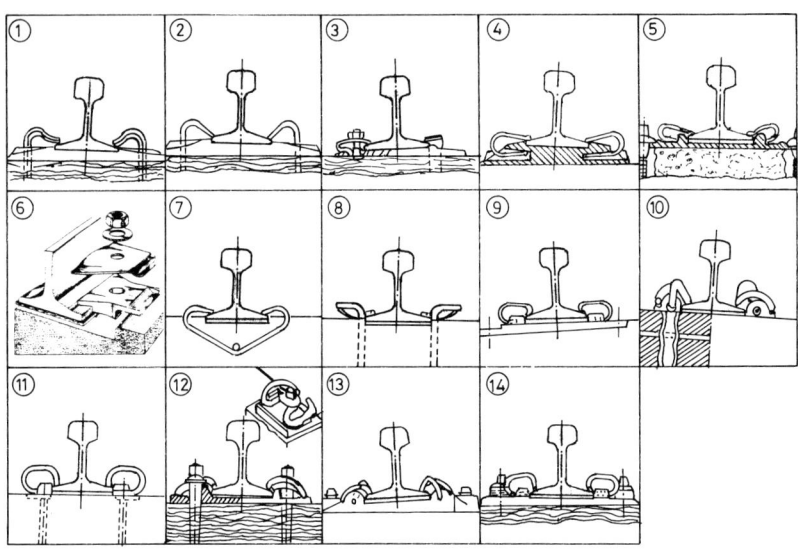

Legend

1	Elastic Spikes	2	Macbeth	3	Rail Anchor	4	Mills
5	Hey-back	6	Nabla	7	Fist	8	Omega (Vossloh)
9	DE	10	Pandrol (without seating plate)	11	Delta (without seating plate)	12	Vossloh
13	Pandrol (with seating plate)	14	Delta (with seating plate)				

6.9.3. Force developed in rigid and in elastic fastenings

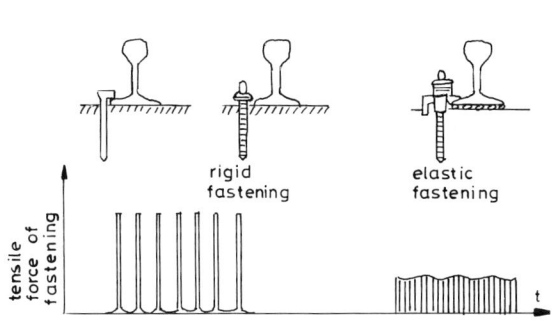

The difference between rigid and elastic fastenings becomes apparent mainly in the diagram of the tensile force developed at the fastening as a function of time (Fig. 6.20). The better behaviour of elastic fastenings is thereby confirmed. Figures 6.21 and 6.22 illustrate the force-elongation curves for screw-type and spring type fastenings respectively.

Fig. 6.20. Force developed in rigid and elastic fastenings as a function of time

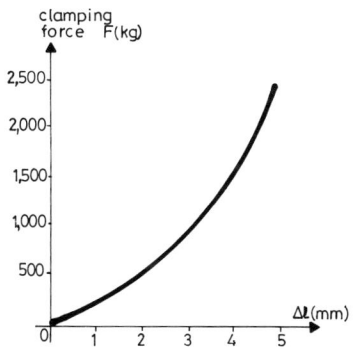

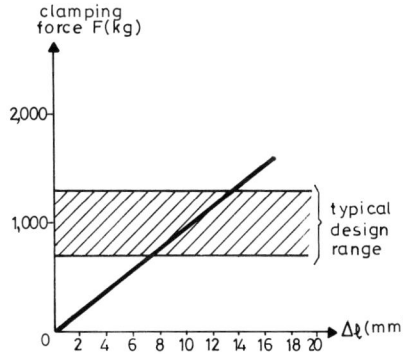

Fig. 6.21. Force-elongation curve for screw-type fastenings

Fig. 6.22. Force-elongation curve for spring-type fastenings

6.9.4. Rail creep and anti-creep anchors

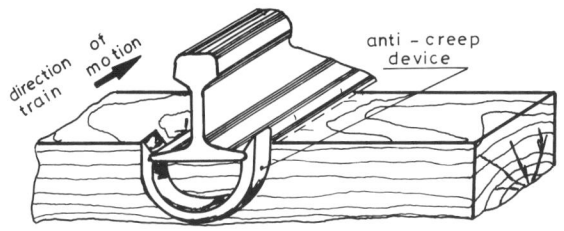

Fig. 6.23. Rail anti-creep device

Along tracks with joints (i.e. not continuous-welded), it is observed that the rails (or even the entire track) are subject to longitudinal creep. Creep usually occurs in the train's running direction. On high gradient tracks, however, rails tend to move downwards, regardless of the direction of traffic. To prevent this slippage, special devices, called anti-creep devices or anchors, are installed along the track (Fig. 6.23).

6.10. Resilient pads

6.10.1. Rail seating and baseplate pads

As explained in chapter 2, section 2.2, resilient pads are used between rail and sleeper or between rail and concrete slab (Fig. 6.24). When a baseplate is used (both in ballasted and non-ballasted tracks), then pads are used between baseplate-sleeper or baseplate-concrete slab and are called baseplate pads (Fig. 6.25).

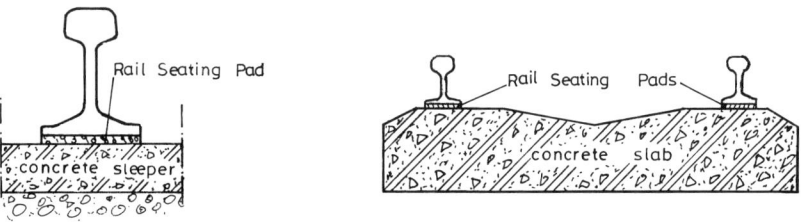

(a) Ballasted track   (b) Non-ballasted track

Fig. 6.24. Rail pads in (a) ballasted and (b) non-ballasted tracks

Sleepers - Fastenings

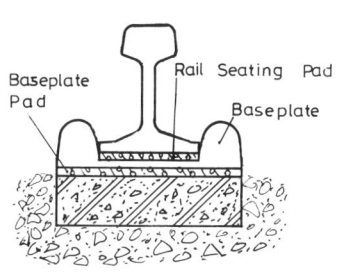
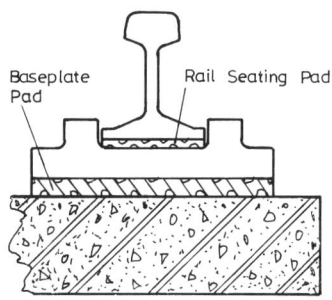

(a) Ballasted track (b) Non-ballasted track

Fig. 6.25. Baseplate and baseplate pads

6.10.2. Functions and properties of pads

Pads must fulfil a number of properties, (144a), (144b):

load distribution. The pad should provide load distribution between the rail foot and the sleeper. It should provide conformable surfaces between rail and sleeper so as to accommodate irregularities on both components

vibration attenuation. The pad should attenuate the transmitted vibrations created by wheel flats and track irregularities from passing trains

resilience. The pad should be designed to provide optimum deflection compatible with the total elastic rail fastening system, such that the fastening is able to provide at all times the necessary resistance to the longitudinal and lateral rail forces. The elastic recovery of pads during train passage should be such that the perceived stiffness at working load is satisfactory

resistance to creep. The pad, together with the rail fastening system, should provide adequate creep and torsonial resistance which should not significantly change with respect to age or tonnage transported

electrical insulation. The pad should have good electrical insulation properties so as to isolate the rails from the sleeper, thus enabling track circuiting to be used for signalling and control purposes

durability. The pad should have a service life of at least as long as the rail. The ideal condition is to install pads during rail replacement. Furthermore, pads should have properties which resist contamination by dirt, water, oil and chemicals, and be able to perform with similar characteristics regardless of ambient temperatures and weather conditions. The Japanese Railways have experienced after 10 years of operation of their Shinkansen high speed train, an increase in the pad stiffness of 66%, (141a).

Railway Engineering

6.10.3. Dimensions and materials

The thickness of the pad (usually 4.5 mm or 9 mm) is chosen to suit the particular installations and depends on several factors:
- the width of the flat-bottomed rail foot
- the type of elastic fastening used
- the size of the sleeper and baseplate, if any
- the type of traffic, e.g. slow speed heavy freight or high speed passenger traffic.

Three main types of materials have been used for pads:
- rubber (both natural and synthetic)
- plastic
- rubber bonded cork.

Thus, French Railways use rubber pads, whilst German Railways use a harder plastic pad. However, certain pads are provided with raised 'pimples', grooves or holes to vary the dynamic and vibration characteristics under load.

6.11. Numerical application for the dimensioning of the various track components

A standard-gauge continuous-welded railway line has a daily traffic load of 30,000 tons, a maximum per-axle load of 20 tons, a maximum speed of 140 km/h, and is laid on medium-quality subgrade (S_2). We will determine:
 a. The most suitable rail type
 b. The most suitable sleeper type. We will examine the cases of timber, twin-block and monoblock sleeper.
 c. The stress distribution under the sleeper.
 d. The most suitable type of fastening.

a. The rail cross-section will be calculated on the basis of the average daily load of 30,000 tons, from Table 5.1. For timber sleepers we select UIC 50 rail, while for twin-block or monoblock concrete sleepers we select UIC 60 rail.

b. Since we have standard-gauge track, the timber sleepers will have the geometric dimensions of Fig. 6.4. If twin-block concrete sleepers are chosen, given that this is a UIC 4 medium-load line (see Fig. 2.6) with a relatively low maximum speed, we will select the sleeper type with the geometric characteristics of Fig. 6.7. Were it a line with a higher load (UIC group 1, 2, 3) and higher speed (V>200 km/h), however, we would have selected twin-block sleepers with the geometric dimensions of Fig. 6.8. In the case of monoblock sleepers a choice of geometrical characteristics can be made based on Tables 6.1, 6.2 and Fig. 6.11.

c. Stress distribution under the sleeper is shown in Fig. 6.14. We will calculate the maximum stress in the case, for instance, of timber sleepers 2.60 m long and 0.15 m wide, (Fig. 6.4).

Sleepers - Fastenings

In order to take into account the dynamic effects (chapter 4, section 4.5), the nominal static per axle load will be multiplied by a factor of 1.5. Given that the sleeper under loading supports only 40% of the axle load (chapter 4, section 4.3.7), the actual load exerted on the sleeper will be:

$$20 \text{ t} \cdot 1.5 \cdot 0.4 = 12 \text{ t}$$

The formula (6.1) (section 6.7) gives:

$$\sigma_1 = \frac{12 \text{ t}}{0.15 \text{ m} \left(\frac{2.60}{2} + \frac{3(2.60 - 1.50)/2}{2}\right)} = 37.65 \text{ t/m}^2 = 3.7 \text{ kp/cm}^2$$

The order of magnitude of stress σ_1 is also confirmed by Fig. 2.3 (chapter 2).

d. In the case of timber sleepers, rigid or elastic fastenings will be selected, while in the case of twin-block and monoblock sleepers, elastic fastenings are indicated and can be selected from Table 6.4.

7 Ballast

7.1. Functions of the ballast and subballast

The term *ballast* in railway engineering denotes the layer of crushed stone (and only in exceptional cases of gravel) on which the sleepers are resting. Furthermore, the ballast fills the space between sleepers as well as some distance beyond the sleeper ends.

The railway ballast (see also chapter 2, Fig. 2.1) performs several functions by:
- further distributing stresses transmitted by the sleepers
- attenuating the greatest part of train vibrations
- resisting track shifting (transverse and longitudinal)
- facilitating rainwater drainage
- allowing track geometry to be restored and correcting track defects (with the use of track maintenance equipment, see chapter 11, section 11.7).

The above functions are clearly contradictory in some aspects, thus the ballast cannot completely fulfil all of them. It could be argued that for good load bearing characteristics and added track stability, the ballast needs to be well graded and compact which, in turn, however, makes dispersal of water more difficult, together with associated maintenance. A balance, therefore, among the various functions that ballast is required to perform is aimed at.

Under the ballast layer is laid the *gravel subballast* that has the following functions:
- protection of the upper surface of the subgrade from the intrusion of ballast stones
- further distributing stresses
- further facilitating rainwater runoff
- imparting a transverse slope (commonly $3 \div 5\%$) to the upper surface of the subgrade for proper runoff.

The usual thickness of the gravel subballast is 15 cm.

Ballast

7.2. Geometrical characteristics of ballast

To fulfil the above functions, ballast must be of good hard stone, angular in shape (cubic or polyhedral) with hard corners, have all dimensions nearly equal and be clean and free from dust.

The ballast shall consist of a mixture of sizes expressed as percentages by weight and evenly graded in any wagon.

Figure 7.1 illustrates a typical granulometric composition of ballast according to the French Railways. Pieces larger than 63 mm and smaller than 16 mm are acceptable up to 3% above and 2% below the limit values. The granulometric composition of ballast according to the British Railways (14 mm ÷ 50 mm) is given in Table 7.1.

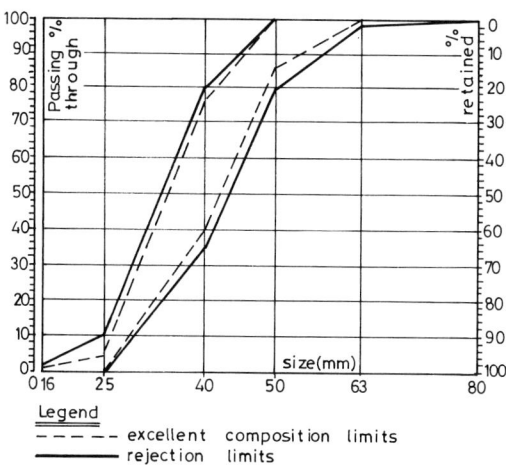

Fig. 7.1. A typical diagram of the granulometric composition of a normal ballast according to the French Railways, (152)

Table 7.1.
Ballast's size according to the
Specifications of the British Railways, (150)

Sieve size D (mm)	Fraction
50 mm	100% to pass
28 mm	less than 20% to pass
14 mm	0% to pass

7.3. Mechanical behaviour of ballast

7.3.1. Stress-strain relationship

On-site measurements of settlements and stresses at the time of passage of train loads have shown that ballast behaviour is elastoplastic, conforming to the Drucker-Prager criterion (see also chapter 4, section 4.3.3.1), (55), (99).

7.3.2. Fatigue behaviour

Both laboratory tests and on-site measurements have shown that on initial loading, the ballast undergoes a considerable permanent (plastic) deformation. In view of its peculiar granulometric composition, the probable cause of this phenomenon is the rearrangement of the stone fragments to attain a state of equilibrium, (145). In subsequent loadings, the contribution of the plastic component to the total deformation is smaller. Accordingly, three-axle testing has shown that the plastic deformation at the n-th loading cycle ε_p^N may be expressed as a function of the plastic deformation at the first loading cycle ε_p^1 by the following expression, (151):

$$\varepsilon_p^N = \varepsilon_p^1 (1 + 0.2 \log N) \tag{7.1}$$

According to equation (7.1), it would take 100,000 loading cycles to double the plastic strain caused by the first loading cycle.

Laboratory tests under constant stress conducted by the British Railways have yielded the following semi-empirical relation for the plastic deformation ε_p^N after N cycles of loading:

$$\varepsilon_p^N = 0.082 \,(100n - 38.2)\,(\sigma_1 - \sigma_3)^\alpha \,(1 + 0.2 \log N) \tag{7.2}$$

where:
  n : ballast porosity
  α : coefficient depending on the level of the stress applied. It ranges from 1 to 2 for low stress values, but may reach 3 for high values of stress.

With respect to the modulus of elasticity, three-axle tests have shown it to change during the first 1,000 loading cycles and to thereafter remain about constant. This is similarly explained as with the appearance of important plastic strain during the first loading cycle. The modulus of elasticity at the one-thousandth loading cycle was found to be about double that at the first cycle, (55), (148).

Ballast

7.4. Ballast hardness

Ballast must have adequate hardness, otherwise it disintegrates and cannot fulfil its functions. Ballast hardness is determined by the Deval and the Los Angeles laboratory tests.

7.4.1. The Deval test

This is the oldest of the tests still in use. It was designed in 1896, at a time when road traffic was composed of carriages with wheels surrounded by steel hoops (tires).

The weight of the test sample (as close to cubic shape as possible) is 5 kg. In the dry test, the sample pieces are washed and dried before being weighed. They are thereafter placed in the cylinders of the Deval machine, which have an internal diameter of 20 cm, an internal length of 34 cm, are inclined by 30°, and are connected to a horizontal axle (Fig. 7.2). The machine is then started (2,000 revolutions per hour) and the entire test takes about 5 hours (a total of 10,000 revolutions).

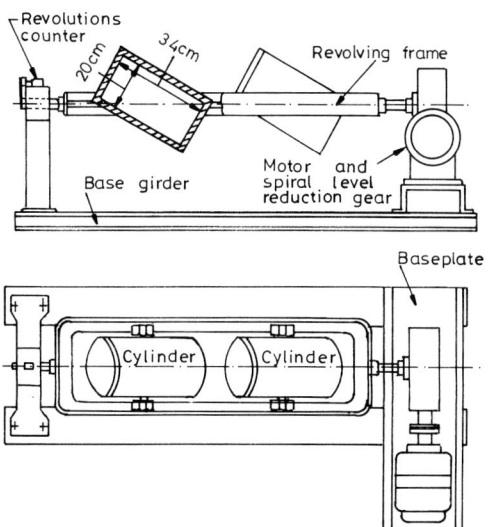

Fig. 7.2. Deval attrition machine

Let A be the initial weight of the sample and B the weight of the sample material retained after the test by a d mm sieve. The value d is according to the French Railways 1.6 mm and according to the British Railways 2.36 mm, (150), (152). Hence, the percentage of attrition will be:

$$W_D = \frac{A - B}{A} \cdot 100 \quad (7.3)$$

The Deval coefficient Q is commonly used and is derived from the relation

$$Q = \frac{40}{W_D} \quad (7.4)$$

French regulations, for instance, specify that the ballast should have a Deval coefficient greater than 14 in the case of hard rock and greater than 12 in the case of limestone rock.

Railway Engineering

The attrition action during the course of the Deval test (the sample has completed 10,000 revolutions at the end of the test) is much stronger than the vibrating action, therefore, only very soft rock is broken to a considerable extent. Pieces with sharp corners in particular are rounded off.

Another variation of the Deval standard test is to carry out the whole procedure in the presence of water, in which case the result is termed as the wet Deval coefficient.

7.4.2. The Los Angeles test

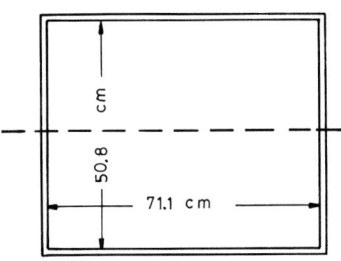

Fig. 7.3. Los Angeles test cylinder

This test is more recent, (1926), than the Deval test. The test equipment consists of a steel cylinder with an internal diameter of 71.1 cm and an internal length of 50.8 cm (Fig. 7.3). A 5 kg sample is placed inside the cylinder together with 12 steel balls, each weighing about 420 g. The cylinder is then set in rotational motion (30 ÷ 33 rounds per minute) until 500 revolutions are completed (duration of the test about 15 minutes).

Let A be the initial weight of the sample and B the weight of the sample material retained after the test by a d mm sieve (d=1.6 mm according to French Regulations and d=2.36 mm according to British Regulations). The percentage of wear is called the Los Angeles coefficient and is:

$$W_{LA} = \frac{A - B}{A} \cdot 100 \qquad (7.5)$$

French regulations specify that the ballast must have a Los Angeles coefficient smaller than 25.

The Los Angeles test has the following advantages:
- action on the inert material is sufficiently strong to bring out any weaknesses,
- it is equally suitable for testing inert materials, crushed rock, and gravel,
- the time required to complete the test is short,
- it reduces the personal influence of the operator to a considerable degree,
- the results of the test agree to a very satisfactory degree with the behaviour of the crushed and inert materials in various construction projects.

Given the aforementioned, the most current technical specifications use the Los Angeles test. Several variations of the Los Angeles test are in existence.

Ballast

7.4.3. Required ballast strength and hardness

The required ballast strength and hardness depend upon the line traffic, the frequency of renewal (usually every 15÷20 years), the nature of the crushed stone, etc. French regulations mandate that the Los Angeles and Deval coefficients intersect at a point lying within the band specified in Fig. 7.4.

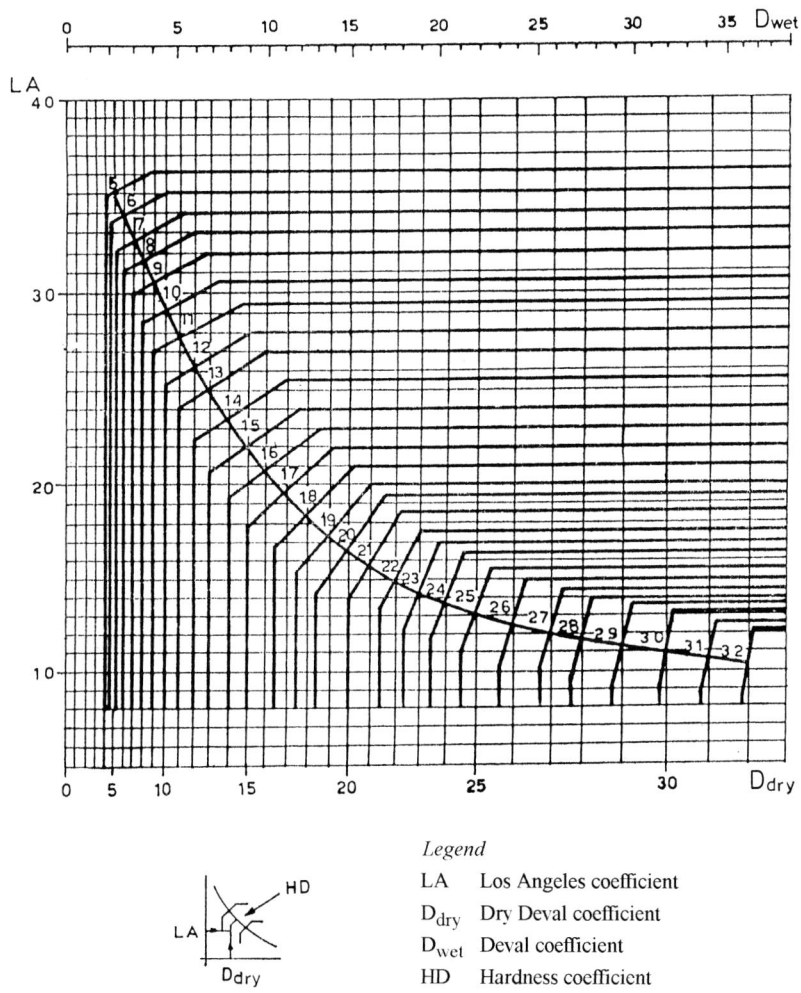

Legend
LA   Los Angeles coefficient
D_{dry} Dry Deval coefficient
D_{wet} Deval coefficient
HD   Hardness coefficient

Fig. 7.4. Combination of the Los Angeles and Deval coefficients for ballast, according to the French Specifications, (152)

Railway Engineering

7.5. Ballast dimensioning

7.5.1. The general dimensioning equation

Until recently, ballast thickness was approximated using nomograms based on the Boussinesq equations. However, finite-element analysis, (91), (142), has allowed all railway parameters to be taken into account:
- line traffic load (see chapter 2, section 2.5.2)
- sleeper material and length (see chapter 6)
- maintenance work volume (see chapter 3, section 3.8, and chapter 11)
- per-axle load (see chapter 2, section 2.5.1).

Ballast thickness, therefore, can be derived from the analytical relation,

$$b(m) = N(m) - a(m) + g(m) - c(m) + d(m) \qquad (7.6)$$

with the following values for each parameter, (142):

Table 7.2.
Values of the parameters used in the ballast dimensioning equation, (142)

Parameter N				
Soil quality of subgrade base	Formation layer			N (m)
	Quality	Thickness e_f (m)		
S_1	S_1	-		0.55
	S_2	$0.30 \div 0.55$		0.40
	S_3	$0.20 \div 0.40$		0.35
S_2	S_2	-		0.40
	S_3	$0.20 \div 0.30$		0.30
S_3	S_3	-		0.30
R (rock)	R	-		0.25

Parameter a(m)
a(m)=0 for lines of UIC groups 1 and 2; 0.05 for lines of UIC group 3; 0.10 for lines of UIC groups 4, 5, and 6, as well as groups 7, 8, 9 with passenger traffic; 0.15 for lines of UIC groups 7, 8, 9 with freight traffic.

Parameter g(m)
g(m) =0 for timber sleepers (L=2.60 m long)
$= \dfrac{2.5 - L}{2}$ for concrete sleepers (L(m): sleeper's length)

Parameter c(m)
c(m) =0 for medium-volume track maintenance works
=0.10 for high-volume track maintenance works

Parameter d(m)
d(m) =0 for per-axle load P = 20 t
=0.05 for per-axle load P = 22.5 t
=0.12 for per-axle load P = 25 t
=0.25 for per-axle load P = 30 t

7.5.2. Numerical application

Application of equation (7.6) with the values of Table 7.2 for a group UIC 4 line (with a yearly traffic ranging from 10 to 18 million tons), with twin-block sleepers (L=2.25 m), per axle load P = 20 t and medium level maintenance gives:
- UIC 4 group ⇒ a = 0.10 m
- twin-block reinforced-concrete sleepers (L=2.25 m long) ⇒ g=0.125 m
- medium-level maintenance ⇒ c=0.05 m
- per-axle load = 20 t ⇒ d=0

Replacing these values in equation (7.6) yields the values of the ballast thickness b which are given in Table 7.3.

Table 7.3.
Ballast thickness b in a group UIC 4 line as a function of subgrade quality

Soil quality of subgrade base	Formation layer quality	Ballast thickness b(m)
S_1	S_1	0.525
	S_2	0.375
	S_3	0.325
S_2	S_2	0.375
	S_3	0.275
S_3	S_3	0.275
R	-	0.225

Ballast thickness is calculated by British Railways in accordance with the speed and tonnage of the line as per Table 7.4 (next page). Values in Tables 7.3 and 7.4 are almost similar, with the exception of a bad quality subgrade (S_1), where an increased ballast depth is needed, when applying formula (7.6), in order to avoid over-stressing and failure of the subgrade.

7.6. Ballast cross-sections

Given the thickness of the various layers, the track cross-sections can be drawn for every case. Various cross-sections are illustrated in next pages for the various types of sleepers. Cross-sections may vary from one network to another because of rolling stock gauge differences, though not significantly, and thus Figs 7.5-7.10 should be used for the determination of the magnitude order of the various layers, always taking into consideration the particularities of each network. The following cases are illustrated:

Table 7.4.
Ballast thickness according to the British Railways

Line speed (km/h)	Yearly line tonnage (million tons)	Ballast thickness b(m)
160 - 200	All	0.38
120 - 160	> 12 million	0.38
120 - 160	2 ÷ 12 million	0.30
120 - 160	< 2 million	0.23
80 - 120	> 12 millions	0.30
80 - 120	< 12 million	0.23
< 80	> 2 million	0.23
< 80	< 2 million (concrete sleepers)	0.20
< 80	< 2 million (timber sleepers)	0.15

- single track with twin-block sleepers - straight track (Fig. 7.5) ⎤
- single track with twin-block sleepers - curved track (Fig. 7.6) ⎬ *non-electrified track*
- single track with steel sleepers (Fig. 7.7) ⎪
- double track with timber sleepers (Fig. 7.8) ⎦
- double track with twin-block concrete sleepers and V_{max} = 300 km/h (Fig. 7.9) ⎤ *electrified track*
- double track with monoblock concrete sleepers and V_{max} = 250 km/h (Fig. 7.10) ⎦

The ballast is superelevated on both sides of the track to increase transverse track resistance, as explained in the following chapter (section 8.4.1).

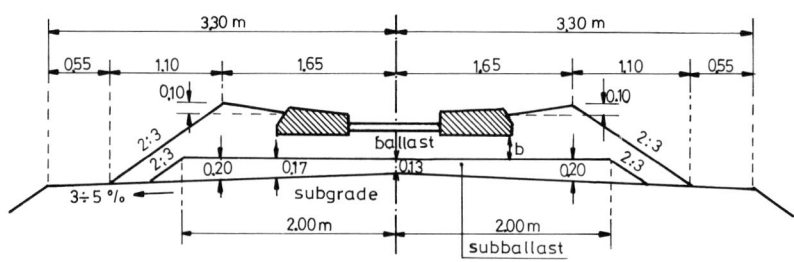

Fig. 7.5. Cross-section of single track with twin-block sleepers (straight track)

Ballast

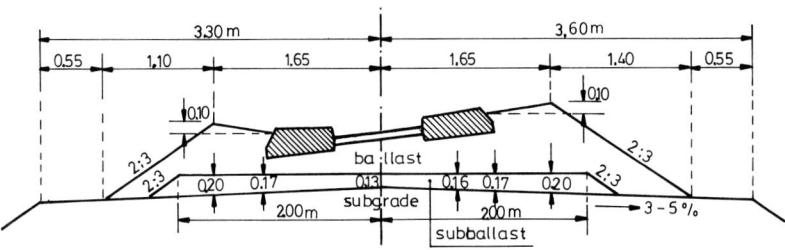

Fig. 7.6. Cross-section of single track with twin-block sleepers (curved track)

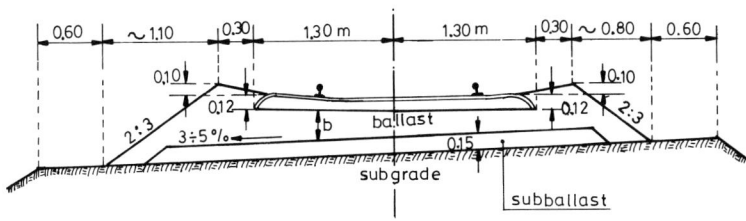

Fig. 7.7. Cross-section of single track with steel sleepers

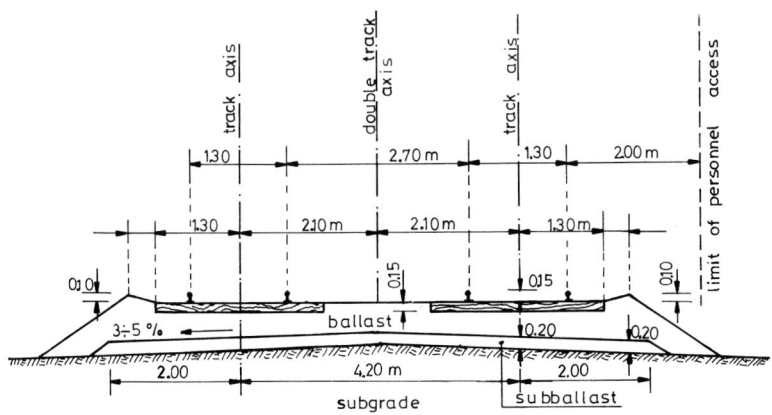

Fig. 7.8. Cross-section of double track with timber sleepers

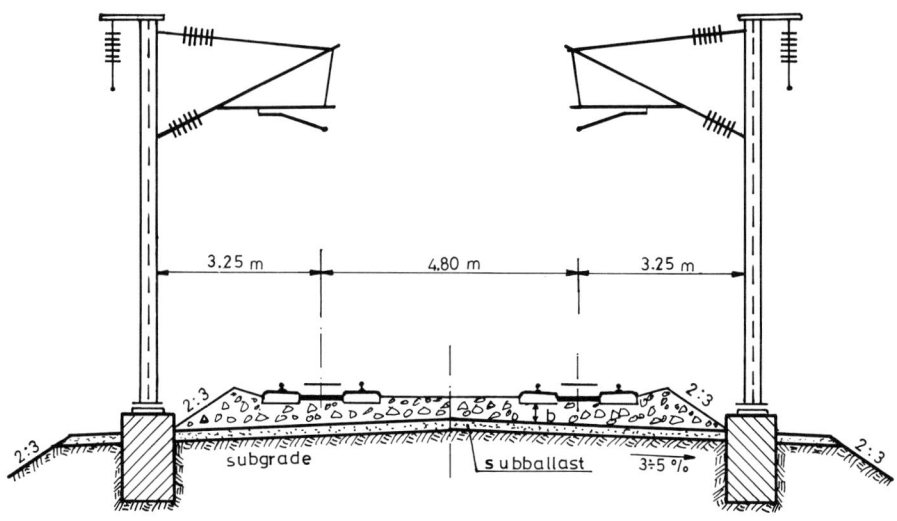

Fig. 7.9. Cross-section of double track with twin-block sleepers and $V_{max} = 300$ km/h

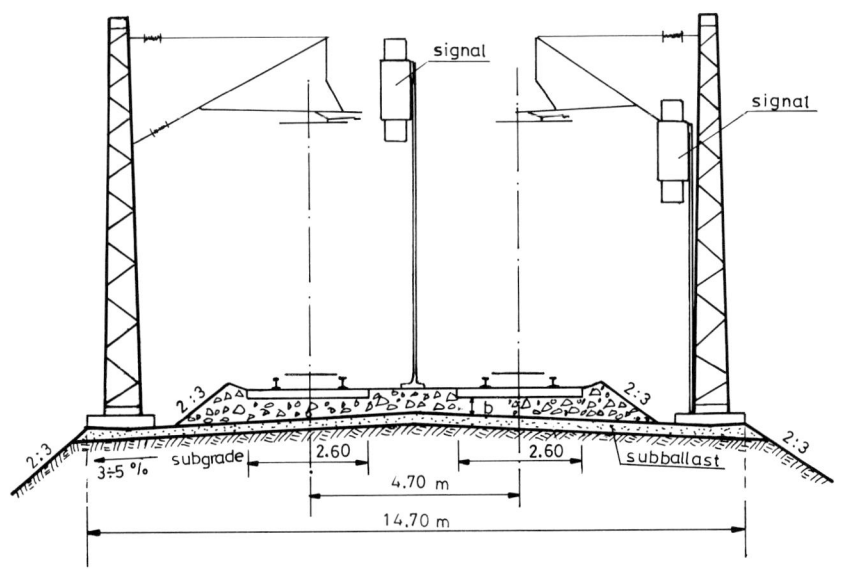

Fig. 7.10. Cross-section of double track with monoblock sleepers and $V_{max} = 250$ km/h

7.7. Laying the track

7.7.1. Mechanical equipment

The laying of the track is carried out, nowadays, with the use in sequence of various mechanical equipment.

Before laying, it should be verified that the subgrade has been properly treated (see chapter 3) and that the transverse slope ($3 \div 5\%$) is correctly given.

Ballast is transported with special wagons and is put in situ. The ballast bed should be properly levelled, graded and in particular consolidated. A gantry or a light type vehicle is used to pull a scarifier for the usual scarifying of the top ballast and also to grade the top ballast with a small grading machine and to consolidate the ballast bed with a vibrating plate or roller vibrator.

Laying of rails and sleepers is done with the use of more sophisticated machines. At present, rails are laid continuous-welded, something that requires careful control of the welding procedures. Another essential feature of rails is cleanliness, that is freedom from oxide inclusions and minimal hydrogen levels. Shelling defects and tache ovales (see chapter 5, section 5.7) should not occur if rails are properly manufactured.

Sleepers should be correctly and uniformly spaced. If groups of sleepers are close together and other groups widely spaced, this will result in an uneven settlement under traffic. The uniformity of spacing is just as important as the nominal spacing.

Pads and fastenings should be properly adjusted on the sleepers. The ideal fastening does not require maintenance, but if it does, then it should be easy and inexpensive.

There are many types of track laying machines. Figure 7.11 illustrates a high speed laying machine. With a daily workshift of 6 hours, this machine achieved an average output of 1.3 km per shift. Peak outputs were 500 m/h and 1.56 km per shift, (150).

Once the track is laid, rails are positioned with the help of a rail positioning machine (Fig. 7.12).

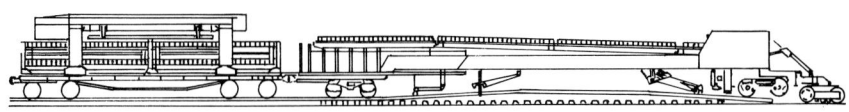

Fig. 7.11. High speed track laying machine

Fig. 7.12. Rail positioning machine

Similar mechanical equipment and methods are used in renewing old tracks, as are used in the case of new tracks on virgin territory. However, additional mechanical equipment is needed for the removal of the old track.

It should be emphasized that fully mechanized track renewal and laying methods would have to be individually adapted to the particular conditions of each railway administration.

However, developing countries with a small railway infrastructure and limited funds, may not find it possible to invest in all the sophisticated material of a fully mechanical laying. For such cases, there is equipment available which enables countries with a surplus of low-cost labour to install modern track assemblies and some of the smaller items of plant together with hand tools such as: sleeper and rail handling tools, manual rail changers, rail skate roller equipment, rail scooters, small hydraulic fastening installation equipment and hand ballast packing machines, rail saw, rail drills, jacks, slewing bars, etc.

7.7.2. Sequence of construction of the various track works

In order to save both labour and time and achieve the best use of the available mechanical equipment, the various trackworks must be well-scheduled. Figure 7.13 gives a bar chart for a typical construction sequence for a single line tracklaying operation with mechanical equipment, (153a).

It can be easily seen that the failure of just one operation, at one location, on one day, will disrupt the whole sequence. Such disruptions should be as few as possible and where a disruption is expected, an alternative sequence of works should be duly reformulated.

Ballasting operations are sometimes put on a low priority on the false assumption that they can always be caught-up afterwards or in slack periods. This has been proven a fatal assumption and many projects are left with many kilometers of unfinished ballasting which in turn affected the final surfacing.

Ballast

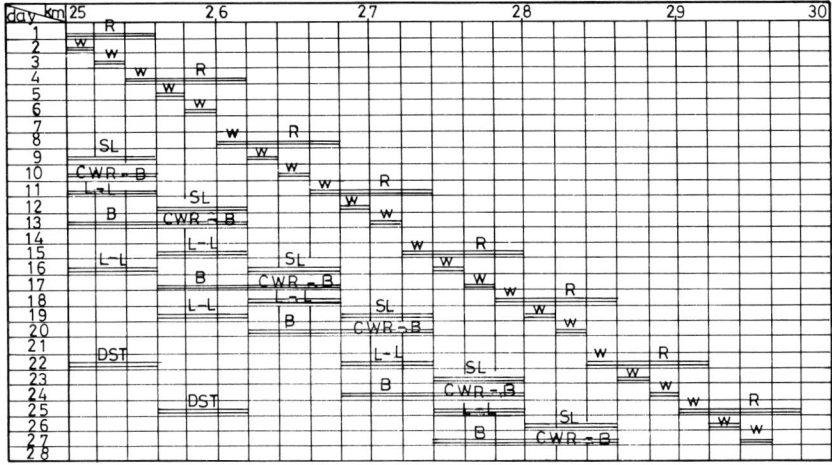

Legend
R Distribute rail   B   Ballast
W Weld   L - L Lift-line
SL Lay sleepers   DST Destress
CWR Install CWR

Fig. 7.13. Track construction program, (153a)

8 Transverse Effects - Derailment

8.1. Transverse effects

During train motion, vertical, transverse and longitudinal forces are applied to the track (see chapter 2, section 2.11.1). Up to this chapter we have mainly examined the influence of vertical forces which determine dimensioning of the various components of the railway track and the subgrade. Transverse forces affect passenger comfort and are crucial to train safety. Exceeding the limits of transverse track resistance may cause track shifting and eventual derailment. Derailment may also be the result of either wheel rebound on the rails or of vehicle overturning, (154). Speed increases in recent years have mandated additional strict protective measures for traffic safety. It should be stressed that compared to other means of transportation, railways are by far the safest.

8.2. Transverse track forces

Let us first examine what transverse forces are applied during train motion. Transverse track forces may be distinguished into two components:
- static
- dynamic

8.2.1. Transverse static force

This is defined as the force due to the unbalanced centrifugal acceleration and to driving forces on curves. Static force $H_s(t)$ will be calculated from the following semi-empirical formula, (157):

$$H_s(t) = \frac{P(t) \cdot NT(mm)}{1,500} \qquad (8.1)$$

where P : per-axle load
NT : transverse defect (see chapter 11, section 11.3.2) if the train is on a straight track, or cant deficiency $h_{d\,max}$ (see chapter 9, section 9.2.2) if the train is on a curve.

8.2.2. Transverse dynamic force

This is defined as the additional dynamic force caused by the various forms of track defects and by rolling stock defects. The dynamic force $H_d(t)$ will be calculated from the following semi-empirical formula, (157):

$$H_d(t) = \frac{P(t) \cdot V(km/h)}{1,000} \qquad (8.2)$$

where P : per-axle load
V : train speed

8.3. Transverse track resistance

The transverse track resistance depends on the sleeper type and on track maintenance. We will consider the worst case, i.e. track condition immediately after maintenance, which destabilizes the track. Under the influence of rail traffic, the ballast is compacted, resulting in an increase of the transverse resistance.

On a track with timber sleepers where maintenance is performed by non-mechanical (manual) means, the transverse resistance may be calculated from the equation, (158):

$$L(t) = 0.85 \left(1 + \frac{P(t)}{3}\right) \qquad (8.3)$$

On tracks with timber sleepers where maintenance is performed mechanically, the transverse resistance is calculated by the formula:

$$L(t) = 1 + \frac{P(t)}{3} \qquad (8.4)$$

On tracks with twin-block reinforced-concrete sleepers, in which maintenance by mechanical means is mandatory, transverse track resistance is:

$$L(t) = 1.5 + \frac{P(t)}{3} \qquad (8.5)$$

Railway Engineering

For tracks with monoblock prestressed-concrete sleepers, small research has been conducted and no analytical formula is available; however, tests have shown that transverse resistance in this case has values between equations (8.4) and (8.5).

The above relations are of a semi-empirical nature and are the result of a series of tests conducted by the French and German Railways, (157), (158). They are currently being used by most railway networks and no objections or reservations have been expressed.

Research on the effects of speed on transverse resistance has shown that the latter is not appreciably affected even at very high speeds, (158).

The above formulas are applicable provided that additional dynamic track loads (see chapter 4, section 4.5) are no greater than 20% of nominal static load. If, however, the additional dynamic loads exceed 20% of static load, the formulas of this section should be multiplied by a correction factor in the order of 0.9, (160). The latter case applies to tracks of medium or bad quality and at rather high speeds.

8.4. Influence of ballast characteristics on transverse track resistance

8.4.1. Influence of the geometrical characteristics of the ballast cross-section

Transverse track resistance is the resultant of the following three components:
* A component generated from friction on the lower surface of the sleeper, proportional to sleeper weight.
* A component resulting from friction between the sleeper sides and the ballast filling the space between consecutive sleepers. This component depends on the degree of filling of the spaces between sleepers (Fig. 8.1), as well as on the degree of ballast compacting. This lateral component amounts to about $40 \div 50\%$ of the total resistance in the case of timber sleepers, $15 \div 25\%$ in the case of twin-block reinforced-concrete sleepers, and 30% in the case of monoblock prestressed-concrete sleepers.

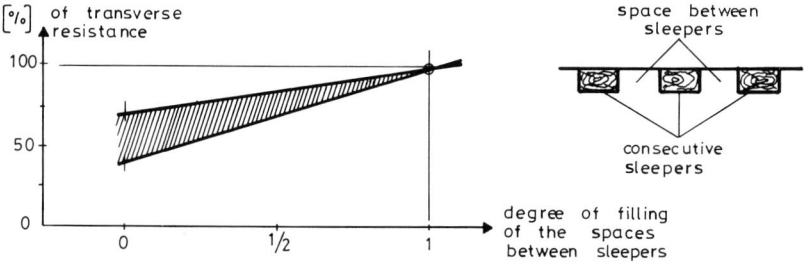

Fig. 8.1. Influence on transverse track resistance of the degree of ballast filling between sleepers

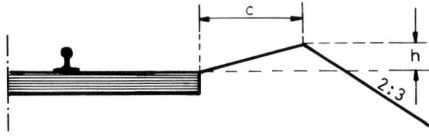

Fig. 8.2. Sleeper end, ballast occupancy width and ballast superelevation

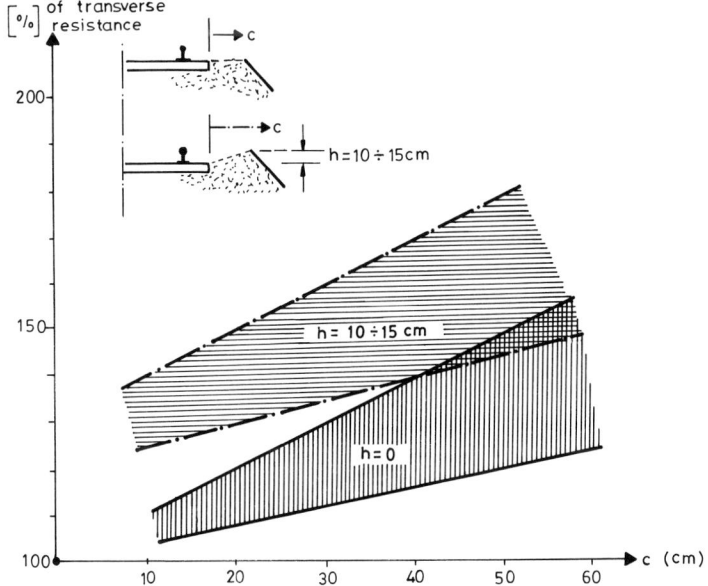

Fig. 8.3. Correlation of transverse track resistance at sleeper end with the geometrical characteristics of the ballast cross-section, (159)

♦ A component developed at the two ends of the sleeper and depending both on the width of ballast occupancy c and whether the ballast is superelevated, (Fig. 8.2).

Figure 8.3 illustrates the increase of transverse resistance caused by an increase of ballast width beyond sleeper ends as well as by superelevation of the ballast section, (159). Therefore, an increase of the ballast cross-section with simultaneous superelevation is preferable to a simple increase of occupancy width.

With respect to the influence of the ballast side slope, its importance is secondary and decreases appreciably as width c is reduced, (159).

Railway Engineering

8.4.2. Influence of the granulometric composition of ballast

The shape and size of the ballast stones, their granulometric composition, and the hardness of the material have a considerable influence on transverse track resistance, (Fig. 8.4).

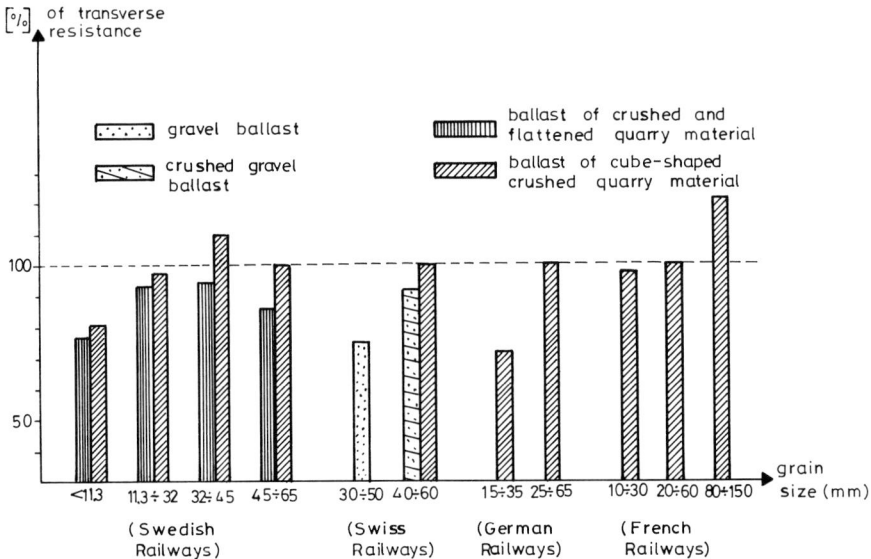

Fig. 8.4. Influence of the granulometric composition of ballast on transverse track resistance, (159)

8.4.3. Influence of the degree of ballast compacting

Following track maintenance works,* the track loses its transverse resistance to a considerable degree (Fig. 8.5). In order to recover transverse resistance, it is necessary to compact the ballast.

Transverse track resistance is fully recovered after the passage of traffic, in particular after the passage of 2 million tons of train loads (Fig. 8.6).

---

* As explained in chapter 11, section 11.7, track maintenance works involve repeatedly raising the track or shifting it horizontally.

Transverse Effects - Derailment

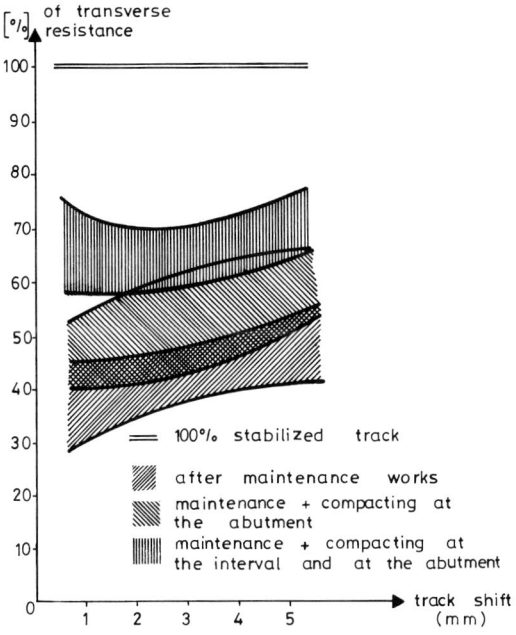

Fig. 8.5. Track stabilization for various forms of compacting

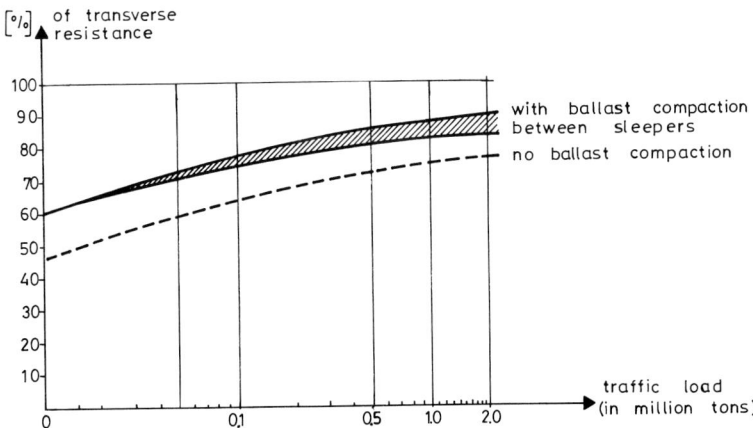

Fig. 8.6. Recovery of transverse track resistance as a function of traffic load

8.5. Influence of sleeper type and characteristics on transverse track resistance

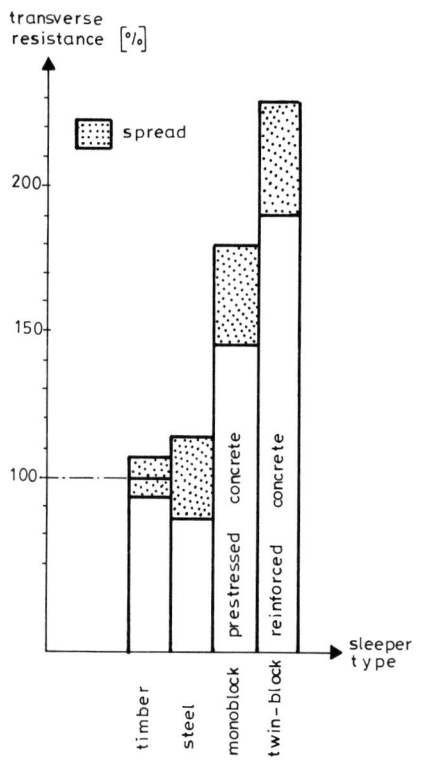

Fig. 8.7. Influence of sleeper type on transverse track resistance, (159)

A series of experimental tests on fully stabilized tracks have shown the unquestionable superiority of concrete sleepers, especially twin-blocks, (159). Figure 8.7 illustrates the transverse resistance for various sleeper types. The relatively large spread is attributable to manufacturing tolerances (dimensions, weight, sleeper form, etc.) as well as to ballast quality and properties.

The high resistance of twin-block sleepers, more than double that of timber sleepers, is mainly due to the following two reasons:

♦ Because of the greater weight of twin-blocks, the resistance component corresponding to the friction between the lower surface of the sleeper and the ballast is greater.

♦ The resistance component generated at the sleeper ends is much greater.

Compared to twin-blocks, the transverse resistance developed in monoblock sleepers is smaller, but clearly higher than that in timber sleepers. This is due to the greater weight, the greater height, and the larger contact surface of monoblock sleepers.

Increasing sleeper length from 2.40 m to 2.60 m in the German Railways has increased transverse resistance by 15 ÷ 20%, (159).

Steel sleeper resistance depends, to a considerable degree, upon sleeper shape (curvature at the ends, ballast contained in the sleeper, etc.). Depending on the sleeper type, the transverse resistance of steel sleepers has values similar to those of timber sleepers.

Concerning timber sleepers, a comparison between the various qualities of wood leads to the following conclusions:

Transverse Effects - Derailment

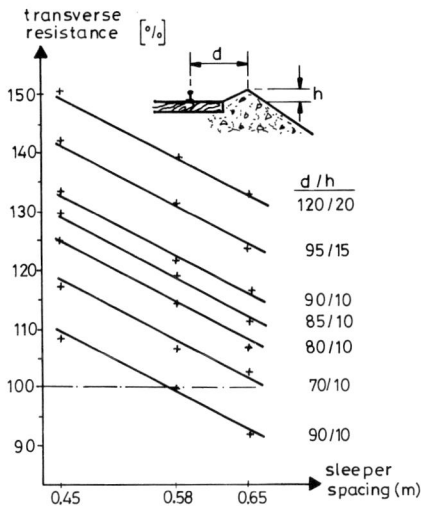

Fig. 8.8. Transverse track resistance as a function of sleeper spacing

Differences between sleepers made of hard wood (e.g. oak) and those made of soft wood (e.g. pine) are minor. Sleepers laid a long time ago, with surfaces roughened by the ballast, especially if the latter has been subjected to compaction, show a transverse resistance slightly higher than new (unused) sleepers.

On the contrary, sleepers made of tropical timber, due to their great hardness and smooth surfaces, have a transverse resistance only 85% that of other qualities of timber, (159).

A reduction of sleeper spacing leads to a slight reduction of the value of per-sleeper resistance, which, however, is more than offset by the greater number of sleepers per kilometre. Overall, transverse track resistance increases when sleeper spacing decreases, (Fig. 8.8).

8.6. Additional measures and special equipment used to increase transverse resistance

In certain cases (e.g. small radii of curvature, turnouts, bridges, etc.) it is necessary to locally increase transverse resistance by special measures not entailing a large expense, such as special sleeper shape, roughened seating surfaces, transverse anchors, etc.

This is a problem of great practical importance, since in certain mountainous areas railway tracks have very small radii of curvature and a high transverse track resistance is necessary due to additional centrifugal loading and internal stresses in rails. Roughening the side and bottom surfaces of timber sleepers improved transverse resistance only slightly. In contrast, cutting grooves into the seating surface of tropical timber sleepers increased transverse resistance by $20 \div 25\%$, (159). The grooves, however, should have sufficient width and depth, so as to ensure that the sleepers grip the ballast well (Fig. 8.9).

Railway Engineering

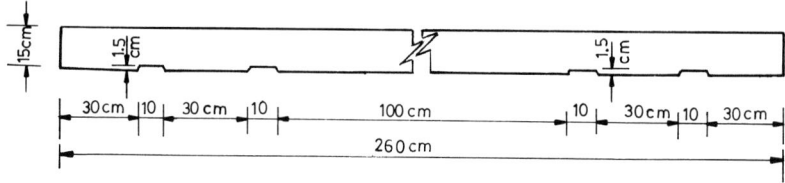

Fig. 8.9. Grooves cut into the seating surface of timber sleepers to increase transverse resistance

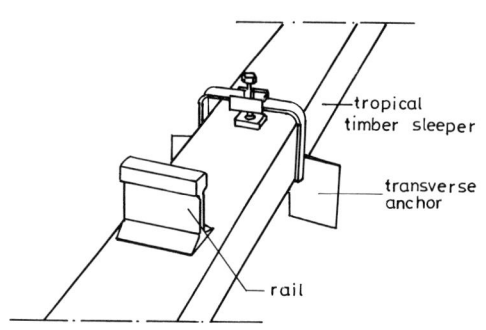

Fig. 8.10. Anchors for transverse resistance

A considerable increase of transverse resistance $(20 \div 80\%)$ is achieved by the so-called transverse anchors (Fig. 8.10), (154). Finally, an even greater increase (in the order of 170%) is attained by placing concrete posts against sleeper ends, but this is an expensive solution interfering with systematic track maintenance.

8.7. Derailment

The derailment of a rail vehicle may occur as a result of one the following:
- Track shifting
- Wheel rebound on the rails
- Vehicle capsizing.

We will discuss each case separately.

8.7.1. Derailment caused by track shifting

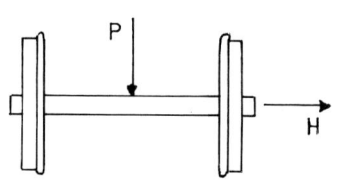

Fig. 8.11. Vertical and transverse forces on a wheel

Under the influence of considerable transverse loading, the track shifts as a whole and causes derailment of the vehicle. This form of derailment mainly occurs at high speeds. The condition for derailment by track shifting is that the transverse force H (Fig. 8.11), which may cause track shifting, exceeds the transverse resistance L, given by formulas (8.3) to (8.5), (section 8.3).

$$H > L \tag{8.6}$$

where

$$H = H_s + H_d \tag{8.7}$$

8.7.2. Derailment caused by wheel rebounding on the rail

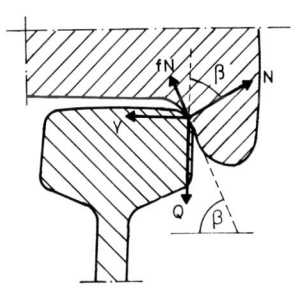

Fig. 8.12. Transverse force between wheel and rail

When the transverse force developed between wheel and rail exceeds a certain value, then the wheel rebounds on the rail and causes derailment. This form of derailment mainly occurs at low speeds and the condition that must be satisfied is given by Nadal's formula (Fig. 8.12):

$$Y = Q \frac{\tan \beta - f}{1 + f \cdot \tan \beta} \tag{8.8}$$

where f is the wheel-rail friction coefficient.

Studies of various cases of derailment have shown that equation (8.8) can be simplified as follows, (107), (157):

vehicle on axles: $\frac{Y}{Q} = 1.2$, vehicle on bogies: $\frac{Y}{Q} = 1.3$ (8.9)

In equation (8.9), Y and Q are the total exerted forces. To the static load Q should therefore be added dynamic loads (see chapter 2, section 2.11.2; chapter 4, section 4.5), which may augment the nominal value of Q (e.g. 10 t/wheel) by up to 50%. With respect to the transverse force Y between wheel and rail, it is of a strongly stochastic* nature and no expression of Y as a function of the rolling stock and track parameters has been formulated to this day. The only way available at present to estimate Y is by on-site measurements on the rail, which, however, are difficult, not very reliable, and of course, the site of a measurement cannot be expected to coincide with a likely derailment site.

Calculation of the force Y may be obtained by considering forces at both rails. In fact, equation (8.8) applies at the outer rail. However, forces at the inner rail may be taken into consideration. In this case we will have:

---

* Stochastic is termed a process which can be approximated only by statistical measurements (e.g. earthquakes). In deterministic processes, in contrast, correlation of cause and effect is possible in advance. Most known processes in mechanics, in spite of the observed spread of the results, are of a deterministic nature (e.g. elasticity, etc.).

$$Y_1 = Q_1 \frac{\tan \beta_1 - f}{1 + f \cdot \tan \beta_1} \quad \text{for the outer rail} \tag{8.10}$$

$$Y_2 = Q_2 \frac{\tan \beta_2 + \tan \gamma_2}{1 - \tan \gamma_2 \cdot \tan \beta_2} \quad \text{for the inner rail} \tag{8.11}$$

$$Y_1 = Y_2 + H \quad \text{equilibrium equation} \tag{8.12}$$

with γ_2 the conical tread.

Equations (8.10), (8.11), (8.12) permit the calculation of transverse forces Y_1, Y_2 at the outer and inner rail.

From a series of derailment accidents, (155), (161), it was derived that the risk of a wheel rebounding on a rail is great when the angle β between wheel and rail is smaller than 32°,

$$\beta_{critical} = 32° \tag{8.13}$$

8.7.3. Derailment caused by overturning of the vehicle

In this case, the vehicle capsizes due to overall unstable equilibrium. It was found that in the worst case (with the centre of mass 2.25 m from the track) for standard-gauge track, a vehicle will overturn when the transverse acceleration reaches $g/3$, (157).

As is explained in chapter 9, section 9.4, tracks are laid for a maximum value of non compensated centrifugal acceleration ranging between $0.5 \div 1.0$ m/sec² and never exceeding a maximum value of 1.0 m/sec² $\simeq g/10$. Therefore, the safety factor against derailment by overturning is $\frac{g}{3} / \frac{g}{10} > 3$.

8.7.4. Derailment safety factor - numerical application

We will investigate the derailment safety factor for a train moving on a curve with the maximum cant deficiency h_d = 100 mm (see chapter 9, section 9.3, Table 9.1) at a speed of $80 \div 120$ km/h. The maximum value of per-axle load is 20 t, the track is laid on twin-block sleepers and is maintained with mechanical equipment. The wheel-rail friction coefficient $f = 0.3$ and the angle β wheel-rail $\beta = 38°$.

a. Derailment by track shifting

According to equation (8.6), derailment by track shifting will occur when transverse track forces exceed transverse resistance, i.e. when $H > L$. Since $H = H_s + H_d$, from equations (8.1) and (8.2) it follows that:

$$H_s(t) = \frac{P(t) \cdot h_{d\,max}(mm)}{1,500} = \frac{20 \cdot 100}{1,500} = 1.33\,t$$

$$H_d(t) = \frac{P(t) \cdot V(km/h)}{1,000} = \frac{20 \cdot 120}{1,000} = 2.4\,t$$

Because derailment takes place on a curve, the parameter NT of equation (8.1) has been taken equal to the limit cant deficiency value $h_{d\,max}$ (see chapter 9, section 9.2.2, equation (9.11) and section 9.4, Table 9.1).

Transverse track resistance is calculated from equation (8.5):

$$L(t) = 1.5 + \frac{P(t)}{3} = 1.5 + \frac{20}{3} = 8.16\,t$$

The safety factor for the particular case and the particular derailment conditions will be

$$v = \frac{L}{H} = \frac{8.16\,t}{1.33\,t + 2.4\,t} = 2.19$$

b. Derailment by wheel rebounding on the track

Wheel rebounding on a rail requires that the value of Y/Q be greater than 1.2 or 1.3 (vehicle on axles or on bogies, respectively). As already explained (section 8.7.2), no analytical expression of Y as a function of track and rolling stock parameters has been found. Therefore, wheel rebounding is considered likely if certain rolling stock characteristics have values different from those specified during preventive maintenance. This form of derailment is predominant at low speeds, especially in the case of empty rail vehicles.

The critical value of angle β wheel-rail has been determined empirically. This critical value is greater in the case of switches and turnouts, because of the lower speeds involved (see chapter 10, section 10.5).

In our case it is:

$$\beta = 38° > \beta_{critical} \simeq 32°$$

and the safety factor for this case of derailment is greater than 1 but smaller than in the previous case.

c. Derailment by capsizing of the vehicle

This form of derailment can be easily studied in conjunction with the geometrical characteristics of the rolling stock. In any case, the safety factor in this case has values greater than 3.

9 Track Layout

9.1. Rail vehicle running on curve and on transition arc

9.1.1. Effects during movement on curve

According to elementary physics, a vehicle running at a speed V on a curve of radius R develops a centrifugal acceleration $\gamma = V^2/R$ and a centrifugal force $F = mV^2/R$ with the following adverse consequences:

- reduction in passenger comfort
- important transverse forces favouring derailment
- increased transverse loading of both track and rolling stock, resulting in considerable wear
- increased vibration.

In order to reduce the above unfavourable effects the following measures are available:
- Using as large a radius of curvature R as possible. Such a measure is not easily implemented, however, due to topographical constraints which often make large radii conditional on expensive civil engineering projects (bridges, tunnels, high embankments or trenches).
- Transverse cant of the external rail in relation to the internal rail, to offset centrifugal forces. Transverse cant greatly decreases transverse effects, without, however, completely counteracting them in most cases, since the transverse cant cannot exceed certain values beyond which rolling stock and track wear become prohibitive.
- Reduction in train speed, which constitutes a last resort solution, since the trend is to increase train speed.

9.1.2. Transition arc - cubic parabola - clothoid

As we know, on a straight line curvature is zero, while on a curve of radius R curvature is 1/R. Therefore, between a straight and a curved track, the curvature changes abruptly from zero to 1/R. This sudden change of curvature is felt by passengers as a jolt.

A variable-radius transition arc, with zero curvature at the beginning and 1/R curvature at the end, is interposed for smooth transition from rectilinear to curvilinear motion.

As a transition curve between a straight line and a curve, a cubic parabola or a clothoid (as in highway engineering) may be used. In railway engineering the curve commonly used by many railway authorities is the *cubic parabola*, in which curvature ρ is (Fig. 9.1):

$$\rho = \frac{1}{R} = -\frac{d^2y}{dx^2} \qquad (9.1)$$

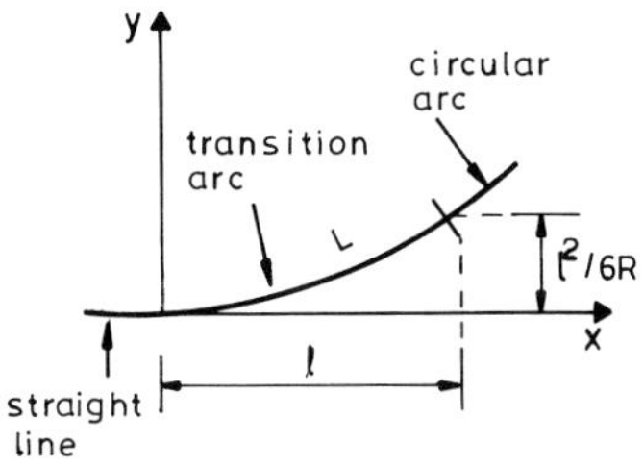

In the cubic parabola, curvature ρ is proportional to the projection of the arc on the x axis

$$\frac{1}{R_{c.b.}} = kx \qquad (9.2)$$

where k is a coefficient.

Fig. 9.1. Cubic parabola

In the cubic parabola it may be assumed that the length L of the arc may be considered equal to its projection ℓ on the x axis. The approximation introduced by this assumption was found satisfactory in most cases.

In the *clothoid* transition curve, the curvature is

$$\frac{1}{R_{cl.}} = kL \qquad (9.3)$$

Using the previous assumption, L = ℓ, it is found that in most cases the use of cubic parabola and of clothoid give similar results.

9.2. Theoretical and actual values of cant - permissible values of transverse acceleration

9.2.1. *Theoretical value of cant for complete compensation of centrifugal forces*

Let us consider a rail vehicle running at a speed V(km/h) on a curve with a radius R(m). We seek the value of the superelevation of the external rail in relation to the inner rail (cant) at which the centrifugal forces are fully compen-

Railway Engineering

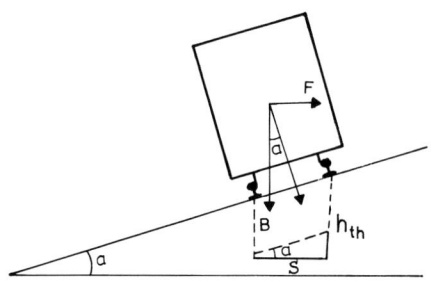

Fig. 9.2. Forces exerted on the rail vehicle and theoretical cant

sated. We will designate this as theoretical cant h_{th}(mm). We have:

$$B = m \cdot g \quad (9.4)$$

$$F = \frac{m \cdot V^2}{R} \quad (9.5)$$

From Fig. 9.2 we have:

$$\tan \alpha = \frac{F}{B} \quad (9.6)$$

as well as

$$\tan \alpha = \frac{h_{th}}{S} \quad (9.7)$$

where
S: spacing of the axes of the two rails, which, in the case of a standard-gauge track, is

$$S = 1,500 \text{ mm} \quad (9.8)$$

From equations (9.4)-(9.8) and after appropriate conversion of units, it is derived for *standard-gauge* tracks that:

$$h_{th}(\text{mm}) = 11.8 \frac{V^2(\text{km/h})}{R \text{ (m)}} \quad (9.9)$$

In the case of *metric-gauge* lines (S = 1,060 mm) it will be:

$$h_{th}(\text{mm}) = 8.3 \frac{V^2(\text{km/h})}{R \text{ (m)}} \quad (9.9.a)$$

9.2.2. Normal or actual value of cant

Equation (9.9) shows that the theoretical value of cant for complete compensation of the centrifugal loads is proportional to vehicle speed. Assuming the latter is constant on a curve, a single value h_{th} of theoretical cant can be calculated. This condition, however, is fulfiled only on metropolitan railways. On the contrary, on conventional railway lines, fast (passenger) and slow (freight) trains coexist.

Thus, if the maximum speed of passenger trains is used in equation (9.9), then passenger comfort is ensured. With freight trains, however, problems arise due to wear of both the wheels and track equipment (particularly of the heads of the internal rails). In particular, if a freight train stops on a curve, it will have trouble starting (it will even be unable to do so if the radius of curvature is too small).

If in equation (9.9) the usual running speed of freight trains is applied, then no problem is created for freight train. Passenger comfort, however, is greatly impaired, as well as the loading of the rail placed higher.

Track Layout

A compromise between the two previous conditions should therefore be found by adopting a cant value ensuring passenger comfort, only moderately increasing rolling stock and track loading, and allowing trains to stop on a curve. This intermediate value of cant h is often termed *normal* or *actual cant*. We will have:

$$h_{th}(V_{min}) < h < h_{th}(V_{max}) \quad (9.10)$$

Selecting the normal value of cant results in cant deficiency for fast trains and cant excess for slow trains.

The difference between the theoretical value of cant for maximum speed and the normal value of cant is termed *cant deficiency* h_d

$$h_d = h_{th}(V_{max}) - h \quad (9.11)$$

The difference between the normal value of cant and the theoretical value of cant for minimum speed is termed *cant excess* h_e

$$h_e = h - h_{th}(V_{min}) \quad (9.12)$$

The normal value of cant, as explained in section 9.4, will be found from the equation:

$$h(mm) = \frac{h_{max}}{h_{max} + h_{d\,max}} \; 11.8 \; \frac{V^2(km/h)}{R(m)} \quad (9.13)$$

In order to deal with the problem of uncompensated centrifugal acceleration, certain types of rolling stock automatically tilt on small-radius curves. With such types of rolling stock, speed can be increased for small radius curves by up to 30% compared to conventional rolling stock. This technique has been applied in the UK, Spain, Sweden, Japan and elsewhere (tilting technology is further analyzed in chapter 12, section 12.11).

9.2.3. Permissible values of transverse acceleration

In chapter 2 (section 2.12) we have seen that passenger comfort depends both on the value of the transverse acceleration and on the duration and frequency at which it is being felt by the human body. The direction in which the transverse acceleration is exerted is also critical. It is found that an acceleration of 0.05 g at a frequency of 1.5 Hz can be tolerated for 5 h 30 min in the vertical direction and 3 h 30 min in the horizontal plane, (1).

Human physiology considerations, therefore, determine the maximum value of transverse acceleration as well as its rate of change. There is general agreement that maximum transverse acceleration should never exceed g/10, i.e. a value of 1 m/sec², (162).

Railway Engineering

In track layout, however, a considerable reduction of passenger comfort cannot be tolerated. Consequently, the uncompensated centrifugal acceleration b should not exceed a percentage of the maximum transverse acceleration γ acceptable by the human body. This limit is set by many railway authorities as follows:

$$b \simeq \frac{2}{3} \gamma \qquad (9.14)$$

In metropolitan railways, a higher value of uncompensated centrifugal acceleration equal to 0.8 m/sec² is considered acceptable.

The acceptable value of b affects the maximum value of cant deficiency.

9.2.4. Cant deficiency variation in time

The variation of cant deficiency in time is:

$$\dot{h}_d \text{ (mm/sec)} = \frac{\Delta h_d}{\Delta t} \qquad (9.15)$$

The parameter $\dot{h}_d$ is expressed as a function of the cant deficiency variation per unit length

$$\dot{h}_d \text{ (mm/sec)} = \frac{\Delta h_d}{\Delta t} = \frac{\Delta h_d}{\Delta \ell} \cdot \frac{\Delta \ell}{\Delta t} = \frac{\Delta h_d}{\Delta t} \frac{V_{max} \text{ (km/h)}}{3.6} \qquad (9.16)$$

9.3. Limit values of cant and acceleration

As will be analyzed in the next section, once values of cant h and acceleration b are defined, then for a given value of speed the radius of curvature R is directly calculated (see equation (9.34) below).

Limit values of cant and acceleration are prescribed by the UIC, (162). Lines are classified in 4 classes:

Class I: V_{max}: 80 ÷ 120 km/h
Class II: V_{max}: 120 ÷ 200 km/h
Class III: V_{max}: 250 km/h, mixed traffic. Standards of German and Swiss Railways are given
Class IV: V_{max}: 300 km/h, only passenger traffic (case of the French TGV).

For each class, normal, maximum and exclusive values of cant and acceleration are given in Table 9.1, (162).

9.4. Calculation of transition curve

In section 9.2.2 we have explained that the value of normal cant h must lie between two limits, to ensure that no problems are caused to either slow (freight)

Table 9.1.
Limit values of cant and acceleration according to the UIC, (162)

Traffic class	I			II			III		III		IV	
Maximum speed V_{max} (km/h)	80 ÷ 120			120 ÷ 200			250 Germany		250 Switzerland		300 France	
Limit value	Norm.	Max.	Excl.	Norm.	Max.	Excl.	Norm.	Max.	Norm.	Max	Norm.	Max
Cant (mm)	150	160	-	120	150	160	65	85	125	-	180	-
Cant deficiency h_d (mm)	80	100	130	100	120	150	40	60	120	-	50	100
Cant excess h_e (mm)	50	70	90	70	90	110	50	70	100	-	-	110
Cant deficiency in time $\dot{h}_d$ (mm/sec)	25	70	90	25	70	-	13	-	36	-	30	75
Non compensated transverse acceleration b (m/sec²)	0.53	0.67	0.86	0.67	0.80	1.00	0.27	0.40	0.81	-	0.33	0.67

or fast (passenger) trains. After the limit values given in the previous section, it should be

$$h_{th}(V_{max}) - h_{d\,max} < h(mm) < h_{th}(V_{min}) + h_{e\,max} \qquad (9.17)$$

and in each case
$$h < h_{max} \qquad (9.18)$$

Selection of a value between the two limits of equation (9.17) depends on the relative density of passenger and freight traffic on the particular line. More passenger traffic raises this value towards the upper limit of equation (9.17), while more freight traffic makes it approach the lower limit of equation (9.17).

In all cases, however, the ratio of the maximum cant h_{max} to the maximum theoretical cant $h_{max} + h_{d\,max}$ should remain constant. Theoretical cant will be multiplied by this constant ratio to find normal cant:

$$h(mm) = \frac{h_{max}}{h_{max} + h_{d\,max}} \; 11.8 \; \frac{V^2 \,(km/h)}{R \,(m)} \qquad (9.19)$$

The minimum value of cant should not result in an uncompensated centrifugal acceleration larger than b_{max}:

$$h_{min}(mm) = 11.8 \, \frac{V^2(km/h)}{R\,(m)} - 152 \, b_{max}(m/sec^2) \qquad (9.20)$$

Railway Engineering

The cant values found from the foregoing equations are rounded off to multiples of 5 mm.

To ensure smooth train running, the value of cant should vary gradually from zero at the end of the straight track to h at the beginning of the circular arc. This requires that the superelevation ramp and the transition curve coincide.

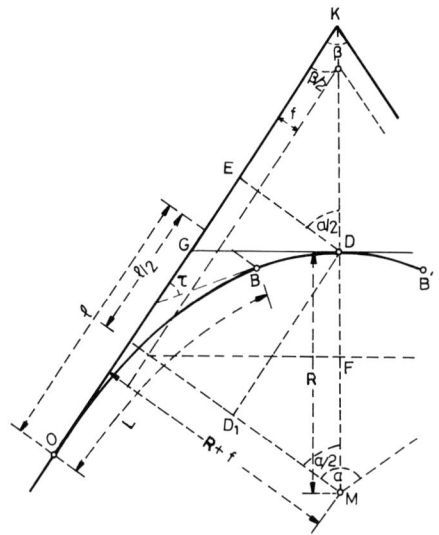

Fig. 9.3. Plan view of a transition (OB: transition arc (cubic parabola); BB′: circular arc)

If L is the length of the transition arc and ℓ its projection on the extension of the straight section, then the minimum value of the transition arc will be

$$\ell_{min}(m) = \frac{h(mm) \cdot V(km/h)}{144} \qquad (9.21)$$

The ordinates of the transition arc, which, as explained, in railway engineering is a cubic parabola, are found from the equation, (166):

$$y = \frac{x^3}{6R\ell}[1 + (\frac{\ell}{2R})^2]^{3/2} \qquad (9.22)$$

In the event that the term $(\frac{\ell}{2R})^2$ is much less than 1, it can be omitted in equation (9.22), in which case we have a small-length cubic parabola. Its equation, applicable as long as $\ell < \frac{R(m)}{3.5}$, is

$$y = \frac{x^3}{6R\ell} \quad (9.23)$$

The ordinates of the cubic parabola are commonly calculated every 10 m, or, whenever a greater point density is required, every 5 m.

The length L of the cubic parabola and its projection ℓ on the straight line are related by the equation:

$$L = \ell + \frac{\ell}{10}\left(\frac{\ell}{2R}\right)^2 \quad (9.24)$$

Certain networks are using parabolic transitions of a higher degree (third or fourth degree parabolas).

Transition curves are not used if:
- the curve radius R > 3,000 m
- the calculated values of cant are practically zero.

9.5. Calculation of the circular arc

Let f be the shift produced by the cubic parabola between straight and circular arc (Fig. 9.3). The characteristics of the circular arc are found from the equations, (8), (166):

$$OK = (R + f) \tan \frac{\alpha}{2} + \frac{\ell}{2} \quad (9.25)$$

$$KD = (R + f)\left(\sec \frac{\alpha}{2} - 1\right) + f \quad (9.26)$$

$$OBD = \frac{1}{2}\left[R \frac{\pi\alpha}{200} + \ell\right] \quad (9.27)$$

where $\sec \frac{\alpha}{2} = \frac{1}{\cos \frac{\alpha}{2}}$ is the secant of the angle $\frac{\alpha}{2}$

and the angle α is expressed in grades.

The shift f is calculated from the formula

$$f = \frac{\ell^2}{24R} \quad (9.28)$$

i.e. in most cases the influence of f on the length OK is negligible compared to the other quantities.

9.6. Case of consecutive same sense and antisense curves

Between two consecutive circular curves of the same sense with radii R_1 and R_2, a transition curve is placed adjacent to each circular curve and an intermediate rectilinear section is interposed between the transition curves. For medium speed tracks (V: $100 \div 200$ km/h), this rectilinear section has a usual value of 30 m.

Using the following symbols

$$\delta = \frac{\ell_2^2}{24R_2} - \frac{\ell_1^2}{24R_1}$$

$$\rho = \frac{R_1 R_2}{R_1 - R_2}$$

$$\ell = \sqrt{24\rho\delta}$$

the transition curve adjacent to the circular curve of radius R_1 will be:

$$y = \frac{x^2}{2R_1} + \frac{\delta}{2} - \frac{1}{6\ell\rho}[(\frac{\ell}{2})^3 - (\frac{\ell}{2} - x)^3] \quad (9.29)$$

Adjacent to the circular curve of radius R_2 the transition curve will be:

$$y = \frac{x^2}{2R_2} + \frac{\delta}{2} - \frac{1}{6\ell\rho}[(\frac{\ell}{2})^3 - (\frac{\ell}{2} - x)^3] \quad (9.30)$$

If interposition of an intermediate rectilinear section is not feasible, then, instead of two transition curves, a single transition curve is used with the following equation:

$$y = \frac{x^3}{6\ell_2 R_2 \cos^3\tau_2} \quad \text{when } L_2 > \frac{R_2}{3.5} \quad (9.31)$$

or

$$y = \frac{x^3}{6\ell_2 R_2} \quad \text{when } L_2 < \frac{R_2}{3.5} \quad (9.32)$$

where L_1, L_2 are the required arc lengths for transition between the rectilinear section and the two circular sections (with radii R_1 and R_2) and τ the angle between the straight line and the tangent at the beginning of the circular arc (Fig. 9.3).

Between two consecutive antisense circular sections are interposed one parabolic transition adjacent to each circular section and an intermediate rectilinear section at least 30 m long (preferably V(km/h)/2). Should the latter not prove feasible, the rectilinear portion is omitted and the two transitions have a common beginning point, a common tangent and the same curvature variation.

9.7. Superelevation ramp

As explained in section 9.4, the superelevation ramp and the cubic transition parabola should coincide. In this case the following cant variation diagram results (Fig. 9.4):

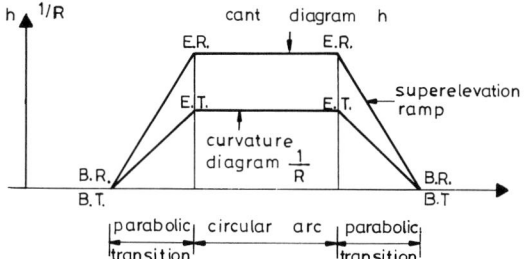

B.R. = Beginning of superelevation ramp
E.R. = End of superelevation ramp
B.T. = Beginning of parabolic transition
E.T. = End of parabolic transition

Fig. 9.4. Cant and curvature variation diagram between rectilinear section and circular curve

A similar linear variation of cant should be applied between same sense or antisense circular curves (Figs. 9.5, 9.6).

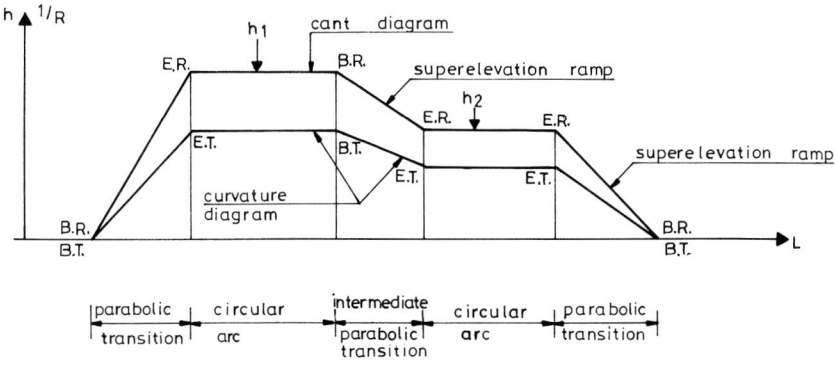

Fig. 9.5. Cant and curvature variation diagram between consecutive same sense circular curves

Railway Engineering

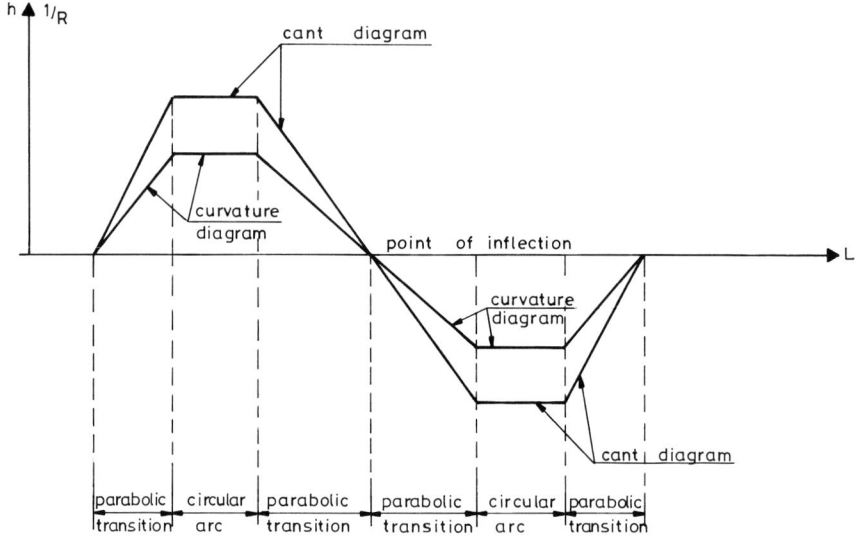

Fig. 9.6. Cant and curvature variation diagram between consecutive antisense circular curves

The maximum gradient ω of the superelevation ramp should not exceed the value $\dfrac{144}{V_{max}}$, i.e.

$$\omega_{max}(\text{mm/m}) < \frac{144}{V_{max}(\text{km/h})} \qquad (9.33)$$

Superelevation ramps should not be located at switches-turnouts or expansion devices. If this is not possible, speed restrictions should be applied.

9.8. Combining maximum and minimum speeds

Equation (9.17) (section 9.4) implies that when maximum and minimum train speeds on a curve differ significantly, it is difficult to find a normal cant value which does not cause problems to freight or passenger trains. A passenger train speed increase is accordingly accompanied by a freight train speed increase, as shown in Table 9.2.

Table 9.2.
Maximum and minimum speed combinations on a curve

$V_{max} < 100$ km/h	→	$V_{min} > 60$ km/h
100 km/h $< V_{max} < 140$ km/h	→	$V_{min} > 70$ km/h
140 km/h $< V_{max} < 200$ km/h	→	$V_{min} > 80$ km/h

9.9. Correlation of the speed on a curve and of its curvature radius

We shall now calculate the maximum permissible speed on a curve of radius R, or, for a given speed V, the minimum permissible radius of curvature.

Obviously, for a given R, the speed V reaches a maximum when the margins for cant h, cant deficiency h_d and cant excess h_e are exhausted.

From equations (9.9), (9.13), (9.17), it follows that

$$\frac{11.8\, V_{max}^2}{R} - h_{d\,max} < 11.8\, \frac{h_{max}}{h_{max} + h_{d\,max}} \frac{V_{max}^2}{R} < \frac{11.8\, V_{min}^2}{R} + h_{e\,max} \quad (9.34)$$

Solving equation (9.34) for V_{max} we have the maximum permitted speed for a given radius R, whereas solving for R we take the minimum required radius for a given speed V_{max}.

With respect to R_{min}, however, it should be ensured that the maximum cant excess for the minimum speed V_{min} of slow trains can be applied. Equation (9.34) gives:

$$\frac{11.8\, V_{max}^2}{R} - h_{d\,max} < \frac{11.8\, V_{min}^2}{R} + h_{e\,max} \quad (9.35)$$

and setting up the maximum values for $h_{d\,max}$, $h_{e\,max}$ and solving for R, we take the minimum radius required by slow trains (with V_{min}).

With respect to the minimum speed, therefore, equations (9.34) and (9.35) should be simultaneously valid, and the higher value found for R_{min} will be used.

Whenever possible, railway networks try to apply the maximum possible value of R. Concerning policy on the lower radius values, there are great differences among networks, principally due to the mountainous or plane character of the ground. Table 9.3 gives the percentage of track curved at 500 m or less in some European networks.

**Table 9.3.
Percentage of curves with a radius smaller than 500 m for various European railway networks (metro systems are not taken into account)**

Country	% of curves with a radius R ≤ 500 m
UK	3.0%
France	9.0%
Germany	13.0%
Switzerland	15.5%
Austria	21.6%

When the radius of curvature of the track is small, track gauge is increased resulting in a value higher than in straight track sections. The increase is applied to the inner rail. For radii R: 300÷600 m, the track gauge can be increased up to 1.455 m (in the case of timber and steel sleepers) and up to 1.440 m (in the case of concrete sleepers), (see also section 11.3.4, table 11.1).

9.10. Gradients

Wherever possible, the longitudinal profile of a line follows the ground profile. Longitudinal gradients of railways are much smaller compared to those of highways. The maximum value of gradient mainly depends on the characteristics and power of the rolling stock. The usual maximum values of gradients on principal lines with mixed traffic and speeds up to 200 km/h range between 12‰÷15‰. The maximum gradient on main lines of German Railways is 12‰, but in the French TGV (with only passenger traffic) it is 35‰ (see also chapter 1, Table 1.8). For adhesion reasons, maximum gradients can hardly exceed the limit value of 40‰. For instance, some lightweight rail systems, which operate vehicles with 50% of the axles motorized, have gradients up to 40‰, (56a). Above this, the use of a rack railway or a funicular must be considered.

9.11. Vertical transition curves

The transition between longitudinal sections with different gradient values is made by interposing a circular arc of radius R_v. The transition is not necessary as long as the difference of the respective gradients (if of the same sense) or their sum (if of opposite sense) is less than 2.5 ‰, i.e. provided that

$$\Delta i < 2.5 ‰ \tag{9.36}$$

The vertical transition radius R_v is calculated from the approximate relation

$$R_v (m) \simeq \frac{V^2 \text{ (km/h)}}{2} \tag{9.37}$$

which in exceptional cases may be reduced to

$$R_{v\,min}(m) \simeq \frac{V^2\,(km/h)}{4} \qquad (9.38)$$

Table 9.4 gives the minimum vertical transition radius as a function of speed.

**Table 9.4.
Longitudinal transition radius as a function of speed**

	Normal value	Exceptional value
V < 100 km/h	5,000 m	2,500 m
100 km/h < V < 150 km/h	10,000 m	5,000 m
150 km/h < V < 200 km/h	20,000 m	10,000 m

The tangent E of the circular transition arc is found from the equation

$$E = \Delta i\,\frac{R_v}{2} \qquad (9.39)$$

where Δi is the gradient difference (Fig. 9.7).

The ordinates of the circular transition arc are calculated from the equation

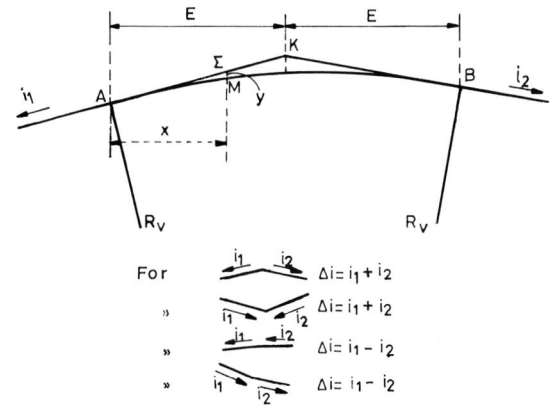

Fig. 9.7.   Vertical transition

$$y = \frac{x^2}{2R_v} \qquad (9.40)$$

No changes of gradient should be made where there are parabolic transitions and hence superelevation ramps exist. Wherever simultaneous vertical and horizontal curves cannot be avoided, the maximum radii of curvature should be used.

Vertical transitions should terminate at least $5 \div 10$ m from the beginning or end of each track switching device. Vertical transitions should moreover be avoided on metallic bridges without ballast.

Railway Engineering

9.12. Layout design using tables

To facilitate layout design, most of the aforementioned equations are used in the form of tables. Such tables spare the designer tedious calculations and give values at a glance. Almost all railway authorities have established such tables many years ago (before the extensive use of computers).

9.13. Layout design using computer methods

Developments in computer hardware and software have revolutionized railway layout design. There are several programs permitting track layout calculation and design requiring only the topography and the limit values of the layout parameters. Figure 9.8 illustrates track layout design for a new line using CAD (Computer Aided Design) software. Furthermore, with computer applications more alternative routes can be easily surveyed and the solution chosen can be studied in more detail.

Fig. 9.8. Track layout design using a computer aided design method

9.14. Construction of a new railway line

9.14.1. Feasibility studies

The decision for realizing a railway project is the outcome of a complex procedure in which politicians, economists and engineers are involved. Feasibility studies are a powerful tool in rationalizing (economically) the choice of the one project to be realized among others.

Feasibility studies compare benefits to costs of the specific railway project.

Cost has two basic components:
- construction cost
- operation cost

Benefits from the realization of a railway project can be, (41)
- reduction of travel time
- reduction of operation cost
- reduction of accidents
- improvement of the quality of service
- regional and national development
- security and national integration, and others.

There are many evaluation methods for transport projects, (166b), (166c).
- In the method of *Present Value (PV)*, all expenses (for construction and operation) are calculated for the lifetime of the project; the alternative with the lowest present value is the most economic one.
- In the method of *Net Present Value (NPV)*, the net present value of each alternative is calculated according to the following formula:

$$NPV = (B - O) - (C - Y) \qquad (9.41)$$

where
- NPV : Net Present Value
- B : Present value of all benefits
- O : Present value of all operation costs
- C : Present value of construction cost
- Y : Residual value (the value of the project at the end of its life)

- In the *Cost-Benefit* method, the ratio λ is calculated as follows:

$$\lambda = \frac{B - O}{C - Y} \qquad (9.42)$$

A project is to be realized if $\lambda > 1$. Among many alternative solutions, the one with the greatest value of λ is chosen.
- In the *Internal Rate of Return (IRR)* method, the value of the discount rate is calculated (by the trial and error procedure), for which the present value of

benefits equals the present value of expenses. If IRR is greater than the opportunity cost of capital, then the specific rail project should be realized, (166b).

Once the decision for realizing a specific rail project is taken, the next step is to conduct the preliminary studies.

In all the above methods, the expected demand after the realization of the new railway project should be forecasted as accurately as possible. Among the many forms of models (time-series models, regression models, gravity models, econometric models, etc), the engineer should choose the most appropriate. At this phase of the study the expected number of passengers or freight traffic, the desired travel times, the areas and urban concentrations that should be served should be established.

9.14.2. Preliminary design

Based on the forecasted demand characteristics, the appropriate available types of rolling stock can be determined. Each rolling stock type is characterized by its power, maximum speed and acceleration, maximum required gradient, etc.

However, the travel times taken into account in the feasibility study determine medium and maximum speeds, which in turn prescribe maximum radii (for layout and vertical transition).

Before beginning the preliminary study, the engineer must collect as much data as possible, which should include the following as a minimum:
- mapping at a scale of 1/50,000 or 1/25,000
- any available aerial photography (eventually from satellites)
- land use and town plans, as well as agricultural plans
- any available geological, hydrological, meteorological and other information
- any previous reports on the study area.

At this preliminary stage, all reasonably possible routes ($2 \div 4$) should be surveyed. For each route, the layout and longitudinal section are studied. The engineer should look for a good vertical profile with as few changes up and down as possible and a good horizontal profile with as few reverses of curvature as possible. Based on these, major technical projects, public utilities to remove and the approximated cost are calculated.

Among the many alternative routes, the best one is chosen with the use of the feasibility methods previously described.

9.14.3. Outline design

Completion of the preliminary design should result in defining a route corridor of interest which may vary from 50 m wide, in a reasonably flat terrain, to maybe 2 km or more in mountainous areas.

The outline design is usually prepared at a scale of 1/5,000 with cross sections surveyed at 100 m intervals.

During this stage, considerations should cover all aspects including the following:
- future traffic and operating demands
- axle load and track gauge parameters
- gradients, minimum radii and other layout characteristics
- subgrade and drainage aspects
- bridges and tunnels
- construction planning

The solution chosen at the end of this phase will be studied in detail in the final design.

9.14.4. Final design

The final stage of the study is generally carried out at scales of 1/2,000 or 1/1,000 in difficult terrain and 1/1,000 or 1/500 in urban areas. Even at this stage of the study, it may take several attempts to attain the right compromise in alignment between speed, curvature, gradient and soil mechanics considerations.

Engineers should always have in mind future maintenance requirements, which should try to minimize and define the appropriate sites and equipment.

9.14.5. Staking of the track layout

After layout calculation and design, its implementation should be preceded by staking. Stakes are driven as follows:
- on double tracks, in the space between the tracks both on straight and on curved sections
- on single tracks, on straight sections regardless of the side of the track (right or left), and in curved sections on the side of the outer rail.

Staking the outer side of the track on curves facilitates the precise laying of the outer rail according to the layout. Alignment of the outer rail is crucial because fast moving trains are guided by the outer rail. The gauge is increased by the suitable positioning of the inner rail.

For economic reasons, double staking is avoided on double tracks and stakes are driven in the space between the two tracks. In this case, the external rail should be laid with great care on curved sections, taking into account the value of the track gauge at the particular point.

On parabolic transitions and circular arcs, stakes are driven every 10 m. Whenever closer staking is necessary, stakes are driven every 5 m.

Along straight sections, stakes are driven every 50 m.

At points where a parabolic transition changes to a straight track, it should be ensured that the extension of the straight section is tangent to the end of the parabola. This is why staking of a parabolic transition is extended by 4 stakes (spaced 10 m) along the straight section to provide at least two zero-deflection points. Alignment of these 4 stakes should be checked by an optical surveying instrument from a point at least 200 m away.

The number of the fixed point and the required cant are marked on each stake.

Staking is followed by the arrangement of the track on the horizontal plane. This stage consists of placing each rail at the proper position on the basis of the fixed points at which the stakes were driven and the values of the track gauge.

10 Switches and Crossings

10.1. Purpose of switches and crossings and their functions

A fundamental characteristic of railways is the one degree of freedom in their movement. In contrast, road vehicles, with two degrees of freedom, can change course easily. In railway engineering, direction is changed by switching devices,* or switches, defined as the equipment and parts through which the direction of movement of a rail vehicle can be changed without interrupting its course.

Switching devices take a great variety of forms. In spite of their apparent complexity, they can be distinguished into two basic forms, and a third, combining the two:
- *simple* (Photo 10.1) or *multiple turnouts*, allowing a track to be split in two (sometimes three) and the moving rail vehicle to change course.

Photo 10.1. Turnout

Photo 10.2. Crossing

---

* Although switches are sometimes referred to as turnouts, the former, strictly speaking, do not include the frogs and check rails enabling one rail to cross the other, while the latter do.

Railway Engineering

- *crossings* (Photo 10.2), where two tracks meet at grade with no change of course
- *turnout crossings*, combining the functions of turnouts and crossings (see below Fig. 10.8, section 10.3).

Thus, the functions of switches and crossings are to enable rail routes to branch from or to join up with one another; to provide flexibility within a route so that trains may move from one track to another parallel track; and finally to enable vehicles to be sorted out. In order to respond efficiently to these requirements, switches and crossings must fulfil a certain number of requirements which include the following:
- impose the fewest possible speed restrictions
- be sited exactly where operational exigencies demand
- provide maximum operational flexibility
- support axle weight required to be carried
- be cheap to manufacture, simple to lay, easily worked, robust and easy to replace
- resist wear, corrosion and decay, and require minimum maintenance
- be compatible with signal engineering requirements

10.2. Arrangement and components of a turnout

In a turnout we distinguish (Fig. 10.1):

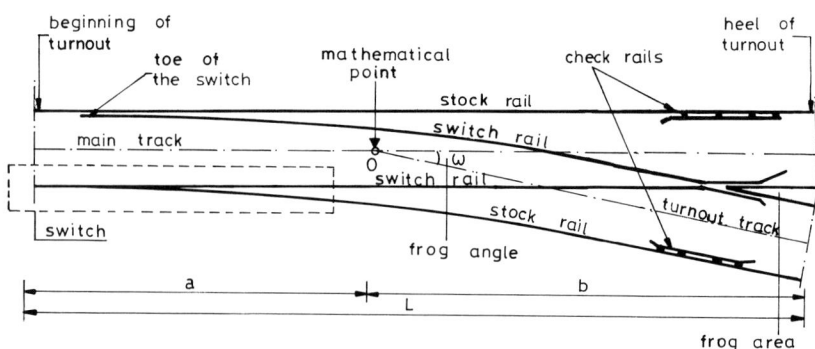

Fig. 10.1. Components of a turnout

- the main track and the turnout track to which the vehicle can be diverted
- the mathematical point O, which is the point where the axes of the two tracks intersect

188

Switches and Crossings

- the frog angle, defined by the axes of the two tracks. The frog angle is commonly denoted by its tangent (e.g. 1:9). The frog angle consists of high-grade material (usually manganese steel)
- the stock rail, which is the rail that stays motionless
- the switch or tongue rail, which is the moving rail section changing the course of the vehicle. A critical parameter is the radius of curvature R of the switch. Depending on their position, switch rails allow rail vehicles to proceed to one or the other track. In switches used in the past, both switch rails were straight. In newer switches, the tongue rail corresponding to the main track is straight, while the other, corresponding to a change of course, is curved.
- the check rail, which is a rail (3 ÷ 10 m long) placed exactly opposite the frog. Shortly before the frog, a wheel reaches a rail gap and it is necessary to provide the other wheel with a guide preventing irregular and uncontrolled movement, which is achieved by installing a check rail. The gap between switch and stock rail is 40 ÷ 130 mm.
- the distances a (from the beginning of the turnout to the mathematical point) and b (from the mathematical point to the end of the turnout)
- the turnout length L (L = a + b)
- the fouling distance c, which is the distance from the beginning of the turnout to the point beyond which a vehicle may stable on one track of the turnout without interfering with the movement of another vehicle on the other track. This point is specified so that the distance between the axes of the two tracks is at least 3.50 m for standard-gauge tracks and 3.00 m for metric-gauge tracks.

Common values for the switch radius R range between 150 ÷ 500 m. For low and medium speed tracks the frog angle (tangent of the angle ω) in old turnouts was given values of 1:8 and 1:10, while in recently installed turnouts it is 1:9 or 1:12, (169).

The cross-section of the switch rail takes form gradually, as shown in Fig. 10.2, (4).

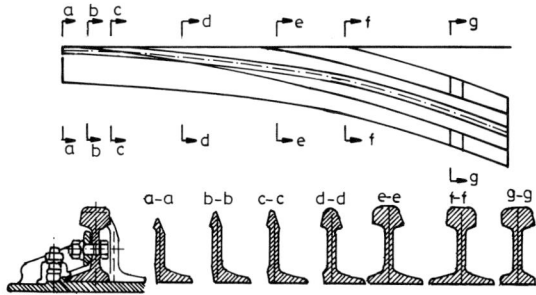

Fig. 10.2. Changing cross-section of the switch rail with increasing distance from the toe of the switch

10.3. Various forms of turnouts

Turnouts and crossings take a great variety of forms depending on the intended change of course. The following are the principal ones.

Fig. 10.3. Standard turnout

- Standard turnout, in which one track is split in two and the main track remains linear (Fig. 10.3).

Fig. 10.4. Simple symmetrical turn-out

- Simple symmetrical turnout, with one track split in two and both the main and the secondary track curving outward (Fig. 10.4).

Fig. 10.5. One-sided double turnout

- One-sided double turnout, with one track successively split into three tracks on the same side and with the main track remaining linear (Fig. 10.5).

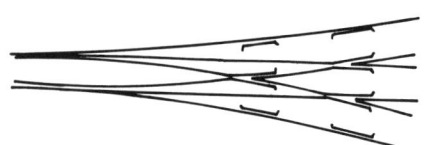

Fig. 10.6. Two-sided double turnout

- Two-sided double turnout, with one track symmetrically split into three tracks: a middle linear track and two symmetrical side tracks (Fig. 10.6).

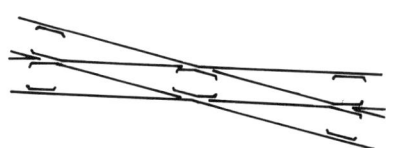

Fig. 10.7. Diamond crossing

- Diamond crossing, where two tracks meet with no change of course (Fig. 10.7).

Switches and Crossings

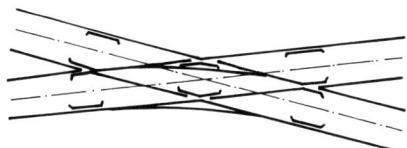

Fig. 10.8. Single slip

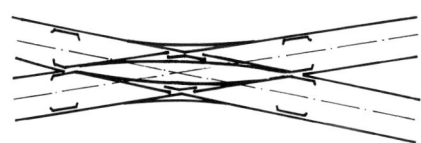

Fig. 10.9. Double slip

Fig. 10.10. Single crossover between two parallel tracks

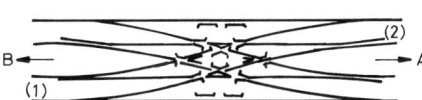

Fig. 10.11. Double crossover between two parallel tracks

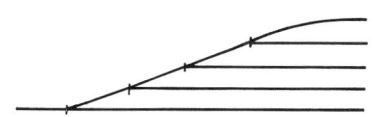

Fig. 10.12. Series of successive turn-outs

• Single slip, where two tracks meet and their course can only be changed from one track to the other in one direction (Fig. 10.8).

• Double slip, where two tracks meet and their course can be changed from one track to the other in both directions (Fig. 10.9).

• Single crossover between two parallel tracks (1), (2). Course can be changed from (1) to (2) in the direction A (or from (2) to (1) in the direction B) but not from (2) to (1) in the direction A (Fig. 10.10).

• Double crossover (sometimes called 'scissors') between two parallel tracks (1), (2). Course can be changed both from (1) to (2) and from (2) to (1) (Fig. 10.11).

• Series of successive turnouts where one track is successively split into several tracks (Fig. 10.12).

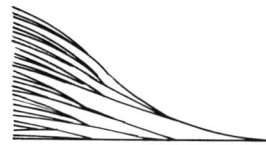

- Track 'fan' with successive track splittings, a technique used in depots and marshalling yards (Fig. 10.13).

Fig. 10.13. Track 'fan'

10.4. Running speed on turnouts

Turnouts differ from regular track in that neither cant nor transition curves are used. Therefore, the maximum running speed on a turnout depends on the value of the uncompensated centrifugal acceleration b and the radius of curvature R of the turnout.

The minimum value of cant in relation to the uncompensated centrifugal acceleration is (see chapter 9, section 9.4, equation (9.20)):

$$h_{min}(mm) = 11.8 \frac{V^2}{R} - 152 \, b_{max} \qquad (10.1)$$

Lateral acceleration b at turnouts must not be too high for reasons of comfort and wear. Usually b_{max} takes values of $0.6 \div 0.7$ m/sec². Thus, for $b_{max} = 0.7$ m/sec², $h_{th} = 0$, we obtain

$$V(km/h) = 3 \sqrt{R(m)} \qquad (10.2)$$

Turnouts are designed as cubic parabolas of small-length (see section 9.4, equation (9.23)) in accordance with the equation

$$y = \frac{x^3}{6 \ell R} \qquad (10.3)$$

From equation (10.2) it is observed that in order to use a train speed of V = 120 km/h, the turnout radius of curvature should be at least R = 1,600 m, while for a speed of V = 150 km/h a radius R = 2,500 m is required.

Such a layout, however, would clearly be excessively extravagant in space requirements. Furthermore it assumes the ability to design a switch in which the tongue rail can be brought truly tangential to the stock rail. For these reasons, in practical turnout design the switch is made much shorter than previous theoretical considerations would demand. This is realized by cutting the stock rail at a finite angle, known as the switch entry angle, (169).

The main characteristics of a turnout usually include its radius of curvature, the frog angle (tangent of the angle ω, see Fig. 10.1), and the tongue rail.

10.5. Derailment protection on switches and crossings

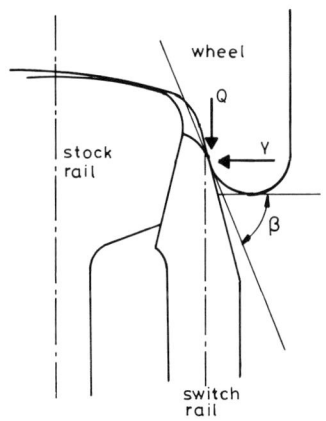

Fig. 10.14. Wheel-rail contact at a turnout

On a turnout or a crossing, a wheel flange may climb a rail causing derailment. To prevent this event, the ratio Y/Q (where Y is the transverse force between wheel and rail and Q is the wheel load) should not exceed a value given by equation (8.8) of section 8.7.2 (Nadal's formula, known also after the names of Boedecher and Chartet, who presented the same formula at the same time with Nadal)

$$\frac{Y}{Q} < \frac{\tan\beta - f}{1 + f \cdot \tan\beta} \quad (10.4)$$

where: β = angle of the wheel flange
f = the wheel-rail coefficient of friction.

Starting at the lowest Y/Q value found from empirical data and the mean value of f, a value preventing derailment can be calculated for the angle β, and therefore the maximum permissible wear of the inner surface of the wheel flange can also be found.

In order to prevent derailment on a turnout, it was derived, (1), (170), that the necessary condition is:

$$\frac{Y}{Q} < 0.4 \quad (10.5)$$

Therefore, for an average value f = 0.3, it is found that

$$\beta_{min} = 40° \quad (10.6)$$

10.6. Switches and crossings for high speeds

Until now it has been assumed that the main line is straight. However, if the main line is curved, the speed at which a switch can be conveniently run will be increased. Let be:

R_o : radius of standard turnout out of straight main-line
R_m : main line radius of curved turnout
R_t : desired radius of turnout of main-line curved at R_m

Railway Engineering

The two arms of a turnout can be in contrary or similar flexure. For contrary flexure:

$$R_t = \frac{R_o \cdot R_m}{R_o - R_m} \quad (10.7)$$

whereas for similar flexure:

$$R_t = \frac{R_o \cdot R_m}{R_o + R_m} \quad (10.8)$$

However, in high speeds the frog angle is reduced. Thus the German Railways use a turnout of a frog angle of 1:42, in which the diverging track can be run at a speed of 200 km/h (lateral acceleration being 0.5 m/sec^2 in this case).

10.7. Sleeper and track layout in turnouts and crossings

In the case of track on twin-block sleepers, timber sleepers are used in the turnout area. If the track is laid on other sleeper types (timber, steel), then the same sleeper type is used for the turnout area as for the remainder of the track.

Sleepers are laid perpendicular to the axis of the main track up to the edge of the check-rail (Fig. 10.15). Beyond this point they are laid perpendicular to the bisectrix of the turnout.

Figure 10.15 illustrates the track and sleeper layout for a turnout type UIC 60 (European specification), while Fig. 10.16 illustrates a turnout according to the American specification.

10.8. Manual and automatic switch operation

A switch may be operated either manually (by local or remote levers) (Photo 10.3) or automatically (Photo 10.4). Automatic operation is driven by electric

Photo 10.3. Manual switch

Photo 10.4. Automatic switch

Switches and Crossings

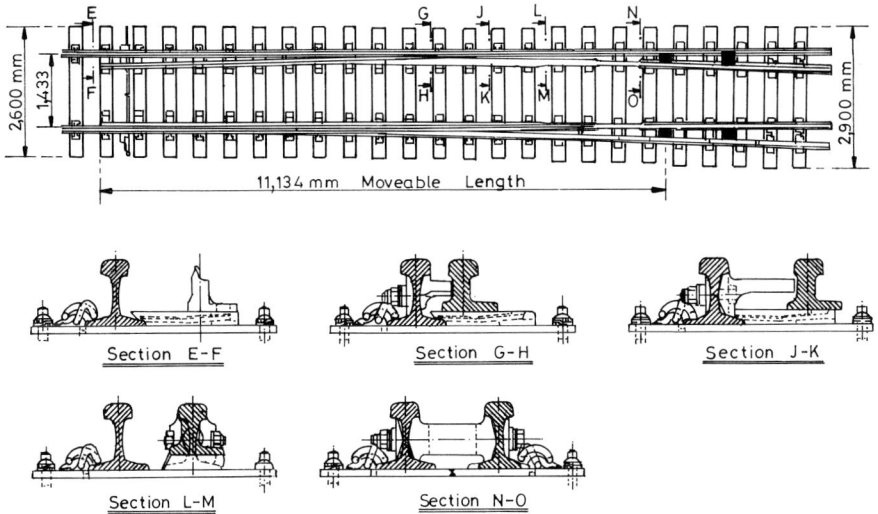

Fig. 10.15. Track and sleeper layout in the case of a turnout type UIC 60 according to the European specification

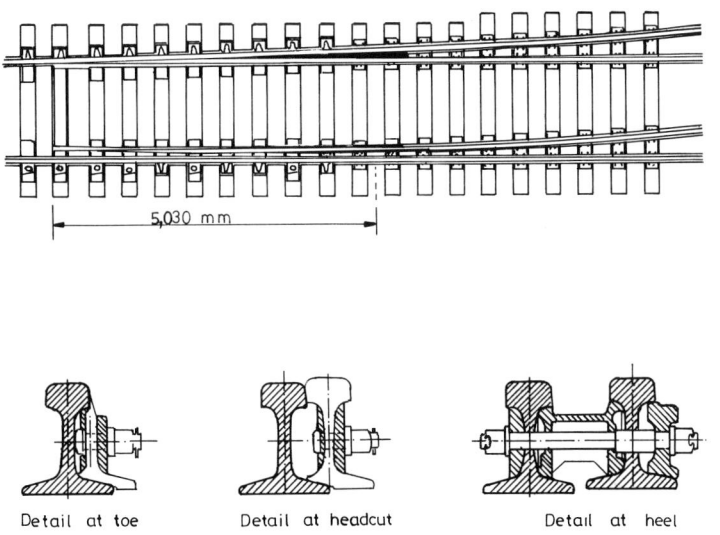

Fig. 10.16. Track and sleeper layout in the case of an American turnout according to the American (AREA) specification

activators operating on commands from electric control boards, operated by station personnel in charge of line traffic. In the case of an automatic switch, the frog triangle is drawn black, while in the case of a manual switch the frog is simply shown in outline.

A switch operates as follows: One of the two switch rails is staying tangent to the rail adjacent to it while the other switch rail leaves from its neighbouring rail a gap sufficient for passage of the wheel flange (Figs 10.1, 10.17). When the set of the two switch rails is operated, either manually or automatically, the above states are interchanged and the switch rail in contact opens, while the other one closes the gap.

In automatic switch operation, the following controls are performed automatically (Fig. 10.17):
- distance between the stock and switch rail
- check rail gauge and wear in the frog area
- setting force, which is the force necessary for reversing the switch device.

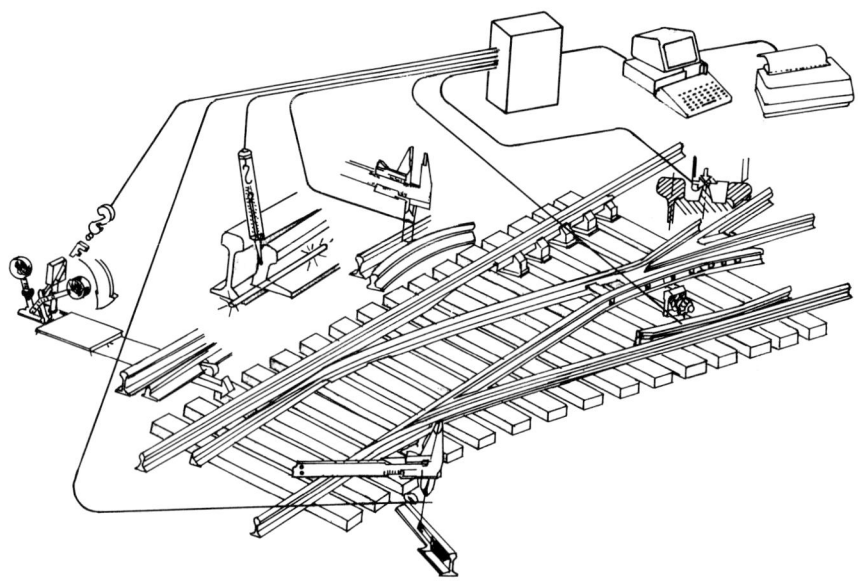

Fig. 10.17. Automatic switch operation

11 Track Maintenance

11.1. Parameters influencing track maintenance

In previous chapters we have examined methods to optimize the design of the track and subgrade and the track layout. However, after the various railway system components start operating, wear begins to appear and, after a certain time, maintenance becomes necessary. Track maintenance decisively affects both train safety and passenger comfort. Track maintenance expenses represent a significant percentage of total railway network expenses.

Therefore, track maintenance expenses should be kept as low as possible while ensuring, for a specific operation speed, that running safety and passenger comfort remain acceptable at all times. With respect to safety, maintenance should be preventive; regarding comfort, maintenance should be corrective; and, finally, as regards the financial aspects of the matter, an optimum solution should be sought, so as to ensure a satisfactory safety margin and prevent an irreparable degradation of track quality.

The above objectives depend on two fundamentally different classes of parameters: on the one hand geometrical parameters, the degradation of which is usually reversible; and on the other, mechanical parameters which in most cases cannot be restored without parts replacement (rails, fastenings, sleepers, welds, etc.).

Geometrical parameters, however, degrade much faster, about $5 \div 15$ times, than mechanical parameters, (172). Accordingly, in lines with an average traffic load ($30,000 \div 40,000$ tons/day, UIC group 4), systematic restoration of geometrical characteristics is done after a load of about $40 \div 50$ million tons, while rails are replaced after about $500 \div 600$ million tons. This means about four years between scheduled maintenance sessions, and rail replacement every $40 \div 50$ years (the above figures are indicative of the order of magnitude only).

Deviations between the actual and theoretical values of geometrical track characteristics are termed *track defects* and their restoration is done through track maintenance. Track defects should be distinguished from *rail defects* (see chapter 5, section 5.7).

Railway Engineering

11.2. Definitions and parameters associated with track defects

Let $z_i(T,x)$ and $z_e(T,x)$ be the elevation of the internal and external rails, respectively, corresponding to a traffic load T (since the last track overhaul), at a kilometric position x. We define the following quantities (Fig. 11.1):

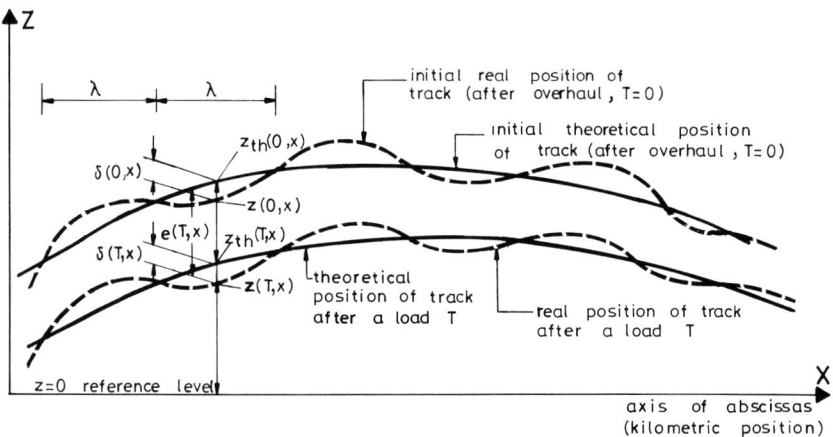

Fig. 11.1. Definition of basic parameters for track maintenance works

Track elevation $z(T,x)$

$$z(T,x) = \frac{z_i(T,x) + z_e(T,x)}{2}$$

Track settlement $e(T,x)$

$$e(T,x) = z(0,x) - z(T,x)$$

Mean settlement $m_e(T)$ **over a track length L**

$$m_e(T) = \frac{1}{L} \int_{x=0}^{x=L} e(T,x)\, dx$$

For measurements performed at discrete positions and not continuously, it will be:

$$m_e(T) = \frac{1}{N} \sum_{i=1}^{N} e(T,x)\, dx$$

Differential settlement $\Delta e(T,x)$

$$\Delta e(T,x) = e(T,x) - m_e(T)$$

Standard deviation of the settlement, sd(T), over a track length L

$$sd(T) = \sqrt{\frac{1}{L} \int_{x=0}^{x=L} [e(T,x_i) - m_e(T)]^2 dx}$$

and for discrete values

$$sd(T) = \sqrt{\frac{1}{N} \sum_{i=1}^{N} [e(T,x_i) - m_e(T)]^2}$$

Theoretical elevation of the track $z_{th}(T,x)$

The real position $z(T,x)$ of the track oscillates around the theoretical position $z^*_{th}(T,x)$ which, being unknown, is approximated over a certain length 2λ around position x, by the value $z_{th}(T,x)$.

$$z_{th}(T,x) = \frac{1}{2\lambda} \int_{x-\lambda}^{x+\lambda} z(T,\xi)\, d\xi$$

11.3. Track defects

11.3.1. Longitudinal defect

The longitudinal defect LD (Fig. 11.2.a) is defined by the equation

$$LD = z_{th}(T,x) - z(T,x) \qquad (11.1)$$

The longitudinal defect is the more reliable in illustrating the effect of the vertical loads on track quality and is a principal factor (together with the transverse defect, see below, which accompanies the longitudinal defect) in determining the magnitude of the track maintenance expenses.

Railway Engineering

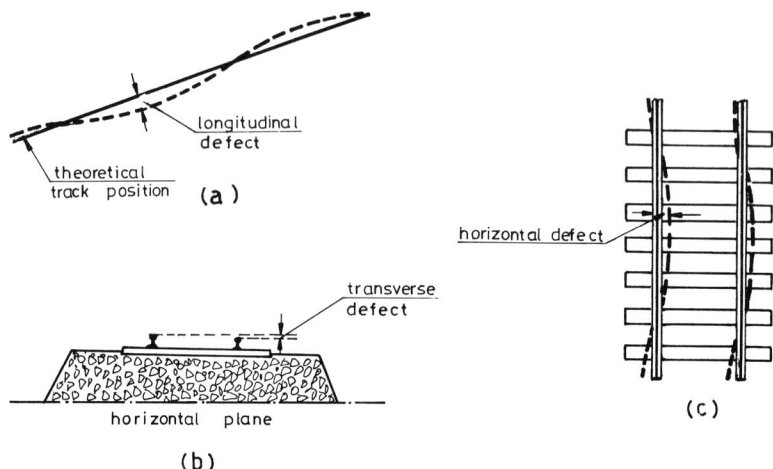

Fig. 11.2. Longitudinal, transverse and horizontal track defects

11.3.2. Transverse defect

The transverse defect TD (Fig. 11.2.b) is defined as the difference between the theoretical and the real value of cant:

$$TD = (z_i - z_e)_{th} - (z_i - z_e) \tag{11.2}$$

11.3.3. Horizontal defect

The horizontal defect HD (Fig. 11.2.c) is defined as the horizontal deviation of the real position of the track from its theoretical position. The horizontal defect depends on the transverse track effects (more than the two previous types of defects) and on the characteristics and particularities of the rolling stock.

11.3.4. Gauge deviations

As explained in chapters 2 and 9, certain track gauge deviations, influenced by the mechanical properties of track materials and the particularities of the rolling stock, are permissible. Gauge values often acceptable for standard-gauge tracks are given in Table 11.1.

Track Maintenance

Table 11.1.
Track gauge as a function of the radius of curvature and sleeper type

Track on timber or steel sleepers		Track on concrete sleepers	
Radius of curvature (m)	Track gauge (mm)	Radius of curvature (m)	Track gauge (mm)
R > 400	1,435	R < 600	1,432
350 < R < 400	1,440	300 < R < 600	1,437
300 < R < 350	1,445		
250 < R < 300	1,450		
R < 250	1,455		

11.3.5. Local distortion

Along straight and circular track sections (where cant is constant), four points of the track lying on two transverse sections (e.g. on two sleepers, as shown in Fig. 11.3) must lie in the same plane. Local distortion ld is defined as the deviation of one point from the plane defined by the other three.

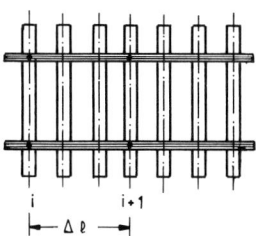

Fig. 11.3. Local distortion: the deviation of one point from the plane defined by the other three

If i and i+1 are two successive transverse sections of the track, spaced Δl apart, (e.g. at the positions of two successive sleepers), then local distortion is defined as the variation of the transverse defect per unit length,

$$ld = \frac{TD_{i+1} - TD_i}{\Delta l}$$

The risk of derailment is prevented when

$$ld < ld_{lim}$$

where ld_{lim} is a value depending mainly on speed and to a lesser degree on the type of the track equipment and of the rolling stock.

It is therefore concluded that the local distortion and the transverse defect are not independent parameters. However, they are often examined separately because local distortion is one of the most frequent causes of derailment, especially at low (V<100 km/h) and medium (V<140 km/h) speeds. The main critical safety parameter at these speeds is local distortion, while other track defects previously mentioned are of lesser importance, (173), (174).

11.4. Track defect recording methods

Track defects were detected until some years ago by competent maintenance personnel either visually (this method, permitting detection of only large defects, did not prove rational nor free of subjective assessment) or by simple instruments. However, in recent years, modern railway technology is using recording vehicles (Fig. 11.4) travelling the track at specified intervals. These vehicles are provided with recording equipment which measure the values of the various track defects in accordance with a specific basis of measurement (in the order of 10 m for longitudinal, transverse and horizontal defects and in the order of $2.5 \div 3$ m for local distortion). Figure 11.5 illustrates a recording of longitudinal defects.

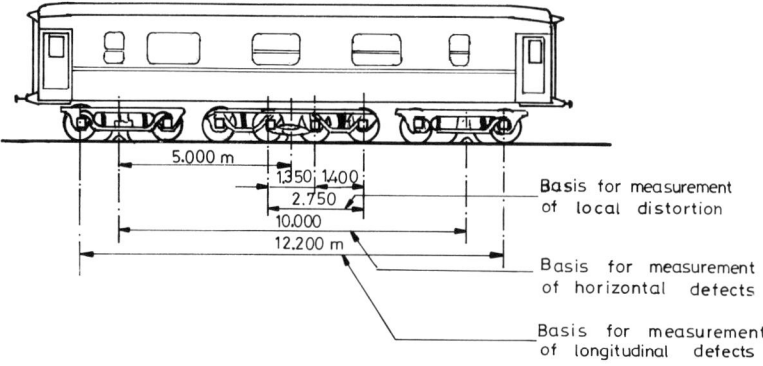

Fig. 11.4. Track defect recording vehicle

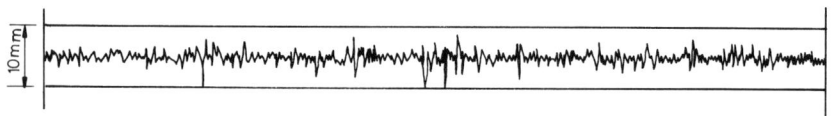

Fig. 11.5. Longitudinal defects as detected by the recording vehicle

The distribution of the various types of defects is of a stochastic nature and can be approximated with the aid of spectral analysis. Thus can be calculated for each class of defects, their frequency of occurrence, the wavelength to which they correspond, their relation to train speed, etc., (4).

A first, and simplest, analysis approach is to calculate the mean (unsigned) values of a defect as well as its maximum discrete values over a particular length. Both the former and the latter will be designated as absolute defect values. They are used at low and medium speeds, where critical and determining parameters are those on which safety depends.

However, at medium, high, and very high speeds, the decisive parameters are those determining passenger comfort. At these speeds, ensuring a high level of passenger comfort also ensures traffic safety. Consequently, as indices of track quality at the above speeds the processed values of the various types of defects are used (obtained from the values recorded by the recording vehicle). Most characteristic of these processed values is the standard deviation of a particular type of defect over a specified length, which reliably measures the differential value of the defect in question, (172).

It should be noted that on medium-speed lines both the absolute and the processed values are used as indices, the former more often.

11.5. Limit values of track defects

For each speed, two limit values are specified:
- *alarm* values of track defects, which, when reached, require the intervention of permanent-way teams on the track. These values will be designated as L_{inf},
- *emergency* values of track defects, which should not be reached, otherwise the deterioration in track quality may become irreversible. Emergency values will be designated as L_{sup}.

The decision for maintenance works should be taken *between the limits L_{inf} and L_{sup}*.

Tracks are usually classified in four categories, depending on train speed, as follows:
I. High speed tracks (V > 200 km/h)
II. Rapid speed tracks (140 km/h < V < 200 km/h)
III. Medium speed tracks (100 km/h < V < 140 km/h)
IV. Low speed tracks (V < 100 km/h), principally lines of UIC groups 7, 8 and 9.

According to the French Railways, the standard deviation for longitudinal and transverse defects and for track categories I, II, III are given in Table 11.2.

For category IV the critical parameter is to avoid derailment, and the emergency absolute value of local distortion should not exceed 4 mm/m, (172).

Table 11.2.
Standard deviation (mm) of longitudinal (LD) and transverse (TD) defects for various categories of tracks, according to the French Railways, (172)

	I		II		III	
	L_{inf}	L_{sup}	L_{inf}	L_{sup}	L_{inf}	L_{sup}
Longitudinal defect LD	0.6	0.8	0.7	1.0	0.8	1.2
Transverse defect TD	0.4	0.6	0.5	0.7	0.6	0.8

11.6. Progress of track defects

The presence of a track defect below the limits specified in the preceding section does not justify intervention by permanent-way teams. Therefore, the question arising is how an initial track defect will evolve as a function of traffic load. Knowledge of the way that track defects evolve may help a timely scheduling of remedial action by permanent-way teams before the limits previously mentioned are exceeded.

11.6.1. Longitudinal defects

A series of tests and statistical analyses, (173), (174), has shown that a defect present in a track after maintenance, progresses rapidly up to a critical load in the order of 2 million tons, beyond which defect progress is much slower. This means that up to this load the track has not fully stabilized, and shows signs of instability.

11.6.1.1. Mean settlement of track

The evolution of the mean settlement is given by the following empirical formula, (173):

$$m_e(T) = a_1 + a_0 \cdot \log \frac{T}{T_r} \tag{11.3}$$

where:
$T_r = 2$ Mt (1 Mt = 10^6 tons)
a_1 : mean settlement for a load T_r (with values of a_1 ranging from $5 \div 15$ mm)
a_0 : settlement increase rate (mm/decade), mainly depending on subgrade quality, with mean values ranging between $2 \div 6$ mm/decade.

Track Maintenance

The ratio $\dfrac{a_o}{a_1}$ illustrates the slow progress of the defect after a load of 2 Mt is reached, and was found to have values ranging between $0.25 \div 0.70$.

$$\frac{a_o}{a_1} = 0.25 \div 0.70 \qquad (11.4)$$

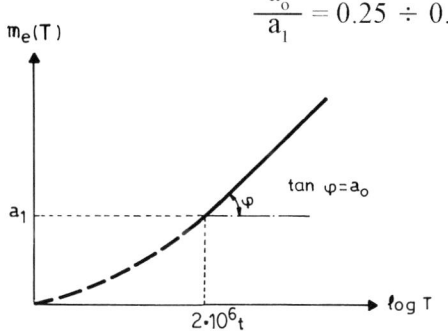

Fig. 11.6. Progress of the mean value $m_e(T)$ of the track settlement as a function of traffic load

11.6.1.2. Standard deviation of the longitudinal defect

For medium and high-speed tracks, the differential value of longitudinal defects LD is of particular interest. Thus, the standard deviation $sd_{LD}(T)$ is accordingly adopted. A series of statistical studies, (173), has yielded the following empirical formula (of a form similar to equation (11.3)):

$$sd_{LD}(T) = c_1 + c_o \cdot \log \frac{T}{T_r} \qquad (11.5)$$

where:
c_1 : standard deviation of longitudinal defects for a load $T_r = 2 \cdot 10^6$ t, with mean values $1.0 \div 1.35$ mm
c_o : rate of increase of standard deviation of longitudinal defects as a function of traffic load, with mean values $0.1 \div 0.2$ mm/decade.

11.6.1.3. Interval between maintenance sessions

Let sd_{LD}^{lim} be the limit value of longitudinal defects, specified by the limits set in section 11.5 (Table 11.2). From equation (11.5) we derive that the limit traffic load T_{lim} between two successive maintenance sessions will be

$$T_{lim} = 2 \cdot 10^6 \cdot 10^{\left(\frac{sd_{LD}^{lim} - c_1}{c_o}\right)} \qquad (11.6)$$

Railway Engineering

Since the quantity c_o is about constant, the determining factors for the time interval T_{lim} between two successive maintenance sessions are the terms sd_{LD}^{lim} and c_1, the latter amounting to the track condition after maintenance. An increase of time T_{lim} is therefore possible by improving the initial condition of the track after maintenance, i.e. by a better quality of track maintenance works.

In the case of medium and low-speed tracks, the average value is used instead of the standard deviation of the longitudinal defect, the form of the above equations remaining the same.

11.6.2. Transverse defect

Transverse defects have a law of evolution similar to equation (11.5). The standard deviation is accordingly given by the following equation:

$$sd_{TD}(T) = u_1 + u_o \cdot \log \frac{T}{T_r} \qquad (11.7)$$

where coefficients u_1 (with mean value 1.2 mm) and u_o (with mean values $0.1 \div 0.4$ mm/decade) are defined similarly to c_1, c_2 of Eq. (11.5).

11.6.3. Horizontal defects

Track loading on the horizontal plane differs from vertical loading in two main respects:
- the resulting effects are much more irregular and discontinuous,
- the stresses developed should, for safety reasons, remain within elasticity limits.

Like other types of defects, horizontal defects progress relatively fast for an initial load T_r in the order of 2 Mt and thereafter slow down considerably. Their law of evolution may be also approximated by a semi-logarithmic formula of the traffic load, which, however, in many instances has shown deviations and a large dispersion. The following equation has been suggested for the mean value of the horizontal defect, (173):

$$m_{HD}(T) = d_1 + d_o \cdot \log \frac{T}{T_r} \qquad (11.8)$$

where coefficients d_1, d_o are defined as in equation (11.3), with mean values $d_1 = 0.6 \div 1.0$ mm and $d_o = 0.15 \div 0.30$ mm/decade.

The ratio $\frac{d_o}{d_1} = 0.2 \div 0.3$, which illustrates how slow the progress of the defects is after the reference load T_r is reached.

11.6.4. Gauge deviations

Gauge deviations mainly depend on subgrade and rolling stock type and therefore their evolution is difficult to determine in terms of the various parameters.

11.6.5. Local distortion

The relation of the evolution of local distortion is also semi-logarithmic:

$$sd_{ld}(T) = g_1 + g_o \cdot \log \frac{T}{T_r} \qquad (11.9)$$

where coefficients g_o (with mean values $1.0 \div 2.0$ mm) and g_1 (with mean values $0.2 \div 1.0$ mm/decade) have a rather large dispersion and are defined as those of equation (11.5).

11.7. Mechanical equipment for maintenance works

Modern railway technology has a panoply of maintenance tools of which the following can be highlighted, (171a), (179), (180):

i) *Heavy* ballast packing, horizontal and elevation correction machinery (photo 11.1, next page), which should be used, to the extent feasible, only in general overhaul operations, where elevation and horizontal adjustment operations are systematic. A necessary condition for their use is that the ballast be sound, free of soil contamination, of proper granulometric size and adequate mechanical strength. The performance of such equipment averages about $200 \div 300$ m per hour.

Underfilling (ballast packing) is the operation whereby track defects are rectified, and includes the following stages:
- A surveying team initially determines the elevation or horizontal correction required to be given the track.
- The ballast packing machine makes a first pass on the track. It moves the track left, right or up depending on the track defects which should be rectified. It lowers the packing blades and compacts the ballast under the sleeper.
- The recording car passes and measures the remaining defects (maintenance tolerances).

To the heavy machinery should also be included:
- Ballast sizing machines, which adjust the ballast cross-section (Photo 11.2).
- Ballast compacting or stabilizing machines, which follow the packing machines and contribute to the increase of the stability and transverse resistance of the track.
- Ballast-clearing machines.

Railway Engineering

Photo 11.1. Heavy ballast-packing machine in the course of track maintenance

Photo 11.2. Ballast sizing machine

ii) *Light* (portable) mechanical ballast-packing machines, the use of which also requires a sound ballast material. Since this equipment is easily transported, it is highly flexible and should be employed in:
- limited operations on discontinuous sections of the track, of up to about 300 m in length, where the use of heavy machinery would be unprofitable,
- repeat packing on particular points of the track,
- elevation adjustment of switches and crossings,
- (as an exception only) for systematic maintenance of track sections, where heavy machinery is not available or is unable to be used on that particular track.

iii) *Hand tools* such as ballast forkand, the pickaxe, etc. which are now considered virtually obsolete, but can still be used in the following cases:
- on track sections with ballast in a state of advanced disintegration where mechanical packing is not possible without adding new sound ballast,
- in the case of isolated, local and urgent repeat packing, where the extent

Track Maintenance

of the operation does not justify the use of even light ballast packing equipment,
- with steel or timber sleepers.

On the opportunity of scheduled maintenance sessions, track equipment is inspected and damages, if any, rectified. For this purpose, the following equipment is used:
- Bolt and screw (both screwing and unscrewing) machines
- Machines for drilling holes into timber sleepers
- Rail cutting machines
- Machines for drilling holes in rails, etc.

11.8. Scheduling maintenance operations

Railways are peculiar in that they consist of discrete subsystems, the interactions of which are neither simple and obvious nor easy to anticipate. Figure 11.7 illustrates a block diagram of the entire maintenance procedure and the

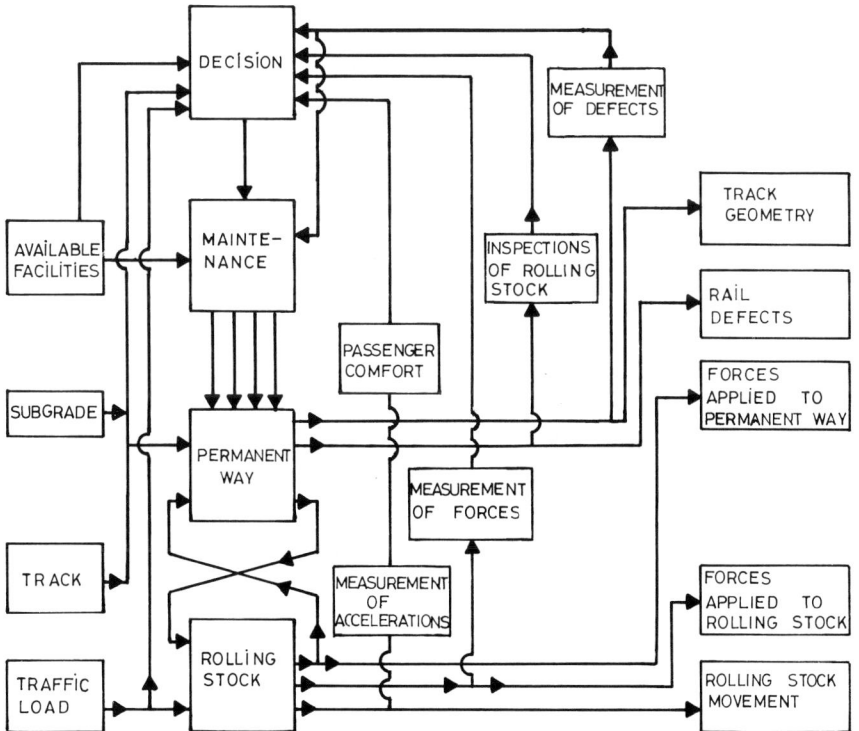

Fig. 11.7. Block diagram of the interactions between the various subsystems and parameters determining maintenance operations

parameters involved. In this chart two processes are apparent, each opposing the other, (171a), (180a):
- The traffic process, which, by the track-rolling stock interaction, tends to increase track defects and to destabilize the system as a whole,
- The maintenance process, which strives to reduce defects and restore the track to its previous good condition.

The two above-mentioned processes should be in equilibrium, which incidentally is the basic purpose of the maintenance works. This equilibrium can be achieved only by timely and rational scheduling, which:
- is based on systematically sorted information from past maintenance operations,
- optimizes the use of the mechanical equipment,
- correctly assigns priorities along the network, on regional and sectional level.

Figure 11.8 (next page) illustrates a flow chart of the successive phases of track maintenance and overhaul. In order to better use both the human and the equipment resources, it is necessary to draw up such diagrams both at the strategic management level and during maintenance works, (178).

11.9. Technical considerations for track maintenance works

When performing maintenance operations, the following should be kept in mind:
- Elevation adjustment is mandatory with any horizontal positioning operation, no later than the next day and in any case before the track stabilizes.
- If elevation adjustment is performed by heavy machinery, horizontal adjustment should be performed simultaneously by the machine.
- If elevation adjustment is performed by heavy machinery, no additional elevation adjustments should be made before track stabilization, which is brought about by line traffic.
- In the event that, after the elapse of the stabilization period, defects not completely rectified are still found, supplementary adjustments should be performed by light equipment, without having to re-lift any elevated sections.

We have seen in the foregoing that after ballast overhaul, a sensitive period follows (until a traffic load of the order of 2 million tons) during which defects evolve rapidly. However, on lines with medium traffic load (groups UIC 4, 5), this period corresponds to about $1 \div 4$ months. On lines with high traffic (groups UIC 1, 2, 3) this period corresponds to $15 \div 40$ days (see also chapter 2, Fig. 2.6). During this time the track should be the object of continuous and careful attention, consisting of the monitoring of the progress of the various defects, and the timely local interventions with light (or heavy, if necessary) machinery,

Track Maintenance

whenever defect accumulation is unusual or excessive. The sensitive period after maintenance is therefore the key to track longevity and to the reduction of future maintenance expenses. If the measures mentioned above are not taken during this period, problems will arise frequently and increased efforts will be required to restore track geometry.

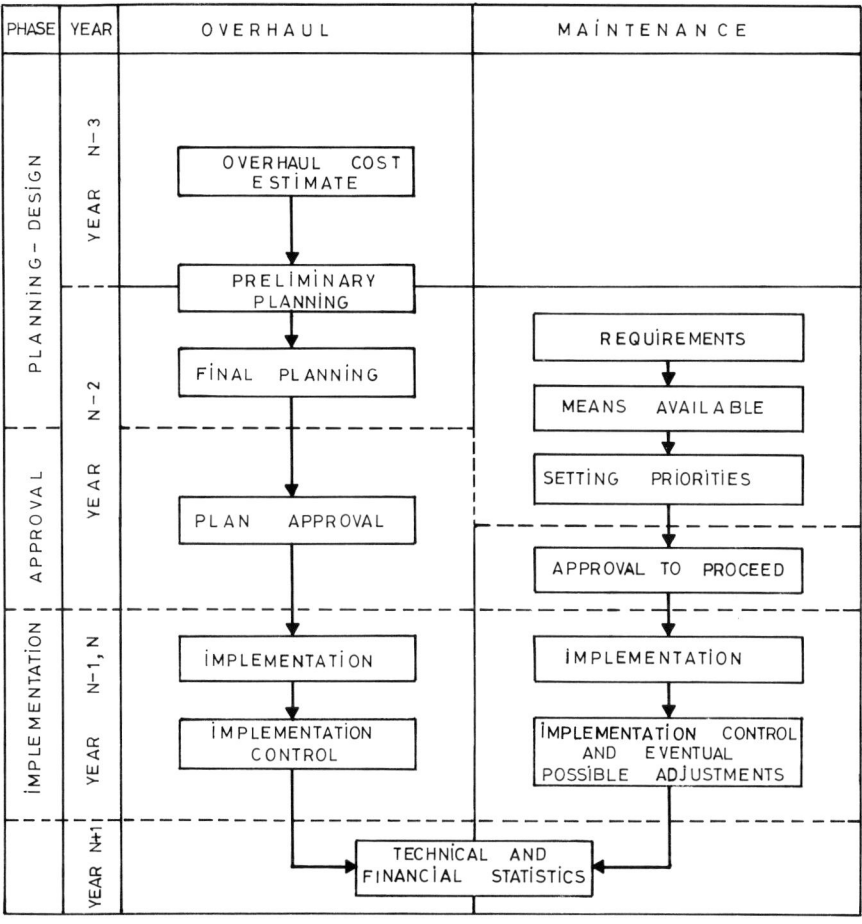

Fig. 11.8. Flow chart of the various planning and execution stages of track overhaul and maintenance operations

11.10. Weed control

Weeds can have serious detrimental effects on the ballast and the subgrade by:
- contaminating the ballast with dirt and vegetation debris, which affects free drainage,
- accelerating the decay of components such as concrete sleepers, not only by chemical action but by the expansion of roots in cracks and crevices,
- by obscuring the track and thus defects normally observed by the naked eye would not be seen on routine visual inspections.

The first railway chemical weed control, introduced in the late 19th century, were arsenic-based chemicals which continued in use in varying degrees in some countries until the 1930s but are not used today. During the 1930s, sodium chlorate was introduced as a chemical weed control and with the addition of fire depressants (such as calcium chloride) was made reasonably safe without any toxic effect. Since World War II, herbicides have been extensively used and particularly hormone selective weedkiller.

Application rates are $1 \div 20$ kg/hectare, the average being $4 \div 8$ kg/hectare. Herbicides must be applied evenly over the area, at the lowest practical volume per hectare if in liquid form, at the greatest practical speed and with maximum safety.

Spray trains can also be used. Their capacity can reach 300 km/day (the daily average observed over a four month season in the UK was 130 km/day), (180b).

12 Train Dynamics

12.1. Train traction

In a train, the locomotive which provides the tractive force is usually distinguishable from these vehicles being pulled. The locomotive may be powered by either internal-combustion (diesel) engines, in which case there is diesel traction, or by electric motors, in which case there is electric traction. The diesel traction is examined in next chapter, section 13.4, and electric traction in sections 13.5 to 13.10.

The pulled vehicles consist of the vehicle body, carrying passengers or freight, and the wheels. The body is supported by the wheels either directly on their axles (see section 12.9.2) or on bogies (section 12.9.3). Wheels which provide traction are referred to as driving wheels, whereas wheels which do not provide traction are known as trailing wheels.

The distinction between pulling and pulled vehicles is less clear in diesel-electric-powered vehicles, where only certain of the otherwise identical passenger vehicles have driving wheels.

In order to ensure train operation at a particular speed, adequate tractive force should be provided to overcome the various forces resisting train motion.

12.2. Forces acting during train motion

12.2.1. The various kinds of forces

During train motion, opposing forces develop which must be overcome by the tractive force. The opposing forces are due to:
- resistance (mechanical and aerodynamic) in horizontal rectilinear motion
- resistance caused by track curves
- a gravity component on gradients, positive when moving uphill, negative downhill
- inertia by acceleration on starting and when speed is not constant.

Railway Engineering

Many of the equations given below are *empirical* or *semi-empirical* and include coefficients with values found for a particular type of rolling stock (e.g. BR: British Railways; DB: German Railways; SNCF: French Railways, etc.).

12.2.2. Running resistances

The force which must be applied to overcome the various resistances of train motion (also applicable to any other land transportation medium) is given by equation (12.1), (181):

$$R = A + B \cdot V + C \cdot V^2 \qquad (12.1)$$

In this equation:
- The terms $A + B \cdot V$ include the various *mechanical* resistances. The first term A (which does not depend on speed, but only on rolling stock characteristics) represents the rolling resistances and those generated by friction between the wheel flange and rail on curves (see section 2.7). The second term $B \cdot V$ represents the various mechanical resistances which are proportional to speed (rotation of axles and shafts, mechanical transmission, braking, etc.).
- The third term $C \cdot V^2$ represents the *aerodynamic* resistances.

The parameters A, B, C can be expressed as functions of the rolling stock characteristics by the following equations (R in daN, V in km/h, other quantities in SI units), (181):

$$A \text{ (daN)} = \lambda \cdot M \sqrt{\frac{10}{m}} \qquad (12.2)$$

where M : total train mass (tons)
   m : mass per axle (tons)
   λ : parameter with values depending on the rolling stock type, e.g. for SNCF vehicles $0.9 < \lambda < 1.5$

$$B \cdot V \text{ (daN)} = 0.01 \, M \cdot V \text{ for good-quality track and} \qquad (12.3)$$
$$\text{rolling stock on bogies}$$

$$C \cdot V^2 \text{ (daN)} = k_1 \cdot S \cdot V^2 + k_2 \cdot p \cdot L \cdot V^2 \qquad (12.4)$$

In equation (12.4), the first term represents the aerodynamic resistances arising at the train front and rear and the second term the aerodynamic resistances generated along the surface $p \cdot L$, where

k_1 : a parameter depending on the shape of the train front and rear. For instance, in conventional SNCF rolling stock, $k_1 = 20 \cdot 10^{-4}$, while for TGV trains $k_1 = 9 \cdot 10^{-4}$, (181)

S : front surface cross-sectional area (in m²) (commonly around 10 m²)

k_2 : parameter depending on the condition of the surface p · L. As an example, in conventional SNCF rolling stock, $k_2 = 30 \cdot 10^{-6}$, while for TGV rolling stock, $k_2 = 20 \cdot 10^{-6}$

p : partial perimeter (in metres) of the rolling stock down to rail level, with common values around 10 m

L : train length (m).

Figure 12.1 illustrates the increase of mechanical and aerodynamic resistances as a function of speed. We see that at high speeds aerodynamic resistance is crucial and trains are given a suitable aerodynamic shape in order to reduce it.

Figure 12.2 illustrates the running resistance as a function of speed as well as the power required to overcome this resistance. We see that in order to increase speed from 200 to 300 km/h, engine power has to be increased by about 200%.

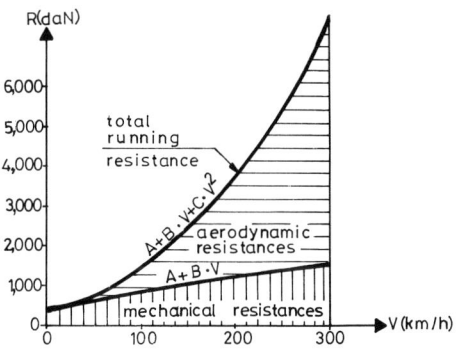

Fig. 12.1. Mechanical and aerodynamic resistances as a function of speed, (23)

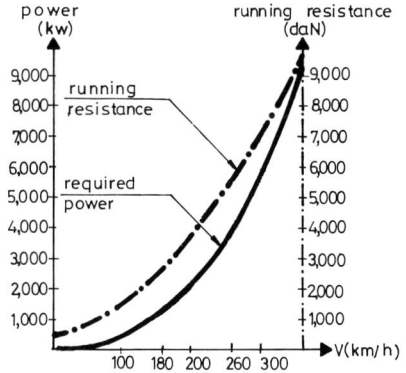

Fig. 12.2. Running resistance and required traction engine power (at zero gradient) as a function of speed (case of the French TGV 001), (22).

Railway Engineering

12.2.3. Empirical formulas for the running resistances

The values of parameters A, B, C of equation (12.1) depend on the characteristics and peculiarities of the rolling stock. The various rolling stock manufacturing firms and the various railway networks have developed empirical formulas to calculate these parameters. Below formulas in use by various railway authorities worldwide are given.

12.2.3.1. Formulas of the French Railways

12.2.3.1.1. Diesel or electric locomotives

The running resistance is given by the empirical relation, (183)

$$R(daN) = 0.65\,L + 13\,n + 0.01\,L \cdot V + 0.03\,V^2 \qquad (12.5)$$

where L : locomotive weight (tons)
       n : number of axles
       V : speed (km/h)

12.2.3.1.2. Pulled rolling stock

Due to the dissimilarity of the pulled rolling stock types, the various formulas present a large spread; they are simplified by merging the term B·V of equation (12.1) with the term C·V². The common practice for pulled rolling stock is to give the resistance per unit weight of rolling stock, also called *specific resistance* r. Therefore, (183):

- For *passenger rail vehicles on bogies*:

$$r(kg/t) = 1.5 + \frac{V^2(km/h)}{4{,}500} \qquad (12.6a)$$

- For standardized *UIC-type* vehicles:

$$r(kg/t) = 1.25 + \frac{V^2(km/h)}{6{,}300} \qquad (12.6b)$$

- For *passenger vehicles on axles and express freight train vehicles*:

$$r(kg/t) = 1.5 + \frac{V^2(km/h)}{2{,}000 \div 2{,}400} \qquad (12.6c)$$

- For *freight vehicles* in the case of a load of 10 t/axle:

Train Dynamics

$$r(kg/t) = 1.5 + \frac{V^2 (km/h)}{1,600} \qquad (12.6d)$$

In the case of 18 t/axle:

$$r(kg/t) = 1.2 + \frac{V^2 (km/h)}{4,000} \qquad (12.6e)$$

12.2.3.1.3. Electric passenger vehicles

Electric passenger vehicles (including traction motors) are commonly used in high-speed trains and in suburban commuter services. The total running resistance R in the case of electric *suburban* trains is given by equation, (183):

$$R(kg) = (1.3 \sqrt{\frac{10}{m}} + 0.01 \text{ V}) \text{ P} + \text{C} \cdot V^2 \qquad (12.7a)$$

where

$$C = 0.0035 \text{ S} + 0.0041 \frac{p \cdot L}{100} + 0.002 \text{ N} \qquad (12.7b)$$

with:
- P : total weight of the electric passenger vehicle (in tons)
- m : weight per axle (tons)
- V : speed (km/h)
- S, p, L : as in Eq. (12.4) (section 12.2.2)
- N : number of raised pantographs (see section 13.9).

In the case of the high speed train *TGV Network* (with a mass of 420 t), total running resistance is given by the formula

$$R(N) = 2,500 + 33 \text{ V (km/h)} + 0.543 \text{ V}^2 \qquad (12.7c)$$

12.2.3.2. Formula of the American Railways

American Railways use the modified Davis formula for the specific rolling resistance, (182):

$$r(lb/t_s) = 0.6 + \frac{20}{M} + 0.01 \text{ V (mph)} + \frac{k}{m \cdot n} V^2 \qquad (12.8)$$

where:
- 1 lb : 0.454 kg
- $1 t_s$: short ton = 2,000 lbs = 907.2 kg
- M : train mass

m : mass per axle
n : number of axles in train
mph : miles per hour = 1.61 km/h
k : C · S
C : air resistance coefficient (from tables)
S : vehicle cross-sectional area (in sq ft)

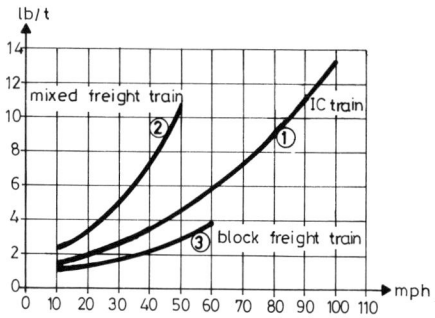

Fig. 12.3. Specific rolling resistance according to the American Railways

Figure 12.3 illustrates the specific rolling resistance for various rolling stock types:

① Intercity train, V = 80 mph, m = 25 lb/axle, train of 16 vehicles with a total mass of 1600 t_s

② Mixed freight train, V = 60 mph, m = 15 lb/axle, average vehicle mass: 45 t_s, total train mass: 3,000 t_s

③ Block freight train, V = 60 mph, m = 60 lb/axle, train of 21 vehicles, each 240 t_s.

12.2.3.3. Formula of the German Railways

The German Railways use for freight trains the Strahl formula (Fig. 12.4)

$$r(Nt/t) = 25 + k \left(\frac{V(km/h) + \Lambda V}{10} \right) \quad (12.9a)$$

and for intercity trains the Sauthoff formula

$$r(Nt/t) = [1 + 0.0025\,V + 0.48 \cdot 1.45\,\frac{16 + 2.7}{800} \cdot \left(\frac{V + 15}{10}\right)^2] \cdot g \quad (12.9b)$$

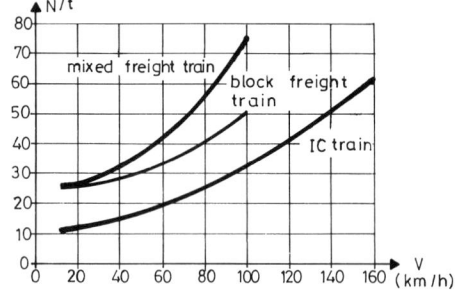

Fig. 12.4. Specific rolling resistance according to the German Railways

where:
k : 0.5 for mixed freight trains and 0.25 for block trains
V : train speed
ΛV : head wind speed (usually taken 15 km/h)

Fig. 12.4 illustrates the specific rolling resistance for various rolling stock types according to the German Railways.

Train Dynamics

12.2.3.4. Formulas for broad and narrow gauge railways

For *broad gauge* (e = 1.676 m), the following formulas have been suggested (Fig. 12.5), (182):

- *passenger trains*
  $$r(Nt/t) = (0.6855 + 0.02112 \, V \, (km/h) + 0.000082 \, V^2) \cdot g \qquad (12.10a)$$
- *freight trains*
  $$r(Nt/t) = (0.87 + 0.0103 \, V + 0.000056 \, V^2) \cdot g \qquad (12.10b)$$

For *narrow gauge* (e = 1.000 m) railways, the following formulas have been suggested (Fig. 12.6), (182):

- *passenger trains*
  $$r(Nt/t) = (1.56 + 0.0075 \, V(km/h) + 0.0003 \, V^2) \cdot g \qquad (12.11a)$$
- *freight trains*
  $$r(Nt/t) = (2.6 + 0.0003 \, V^2) \cdot g \qquad (12.11b)$$

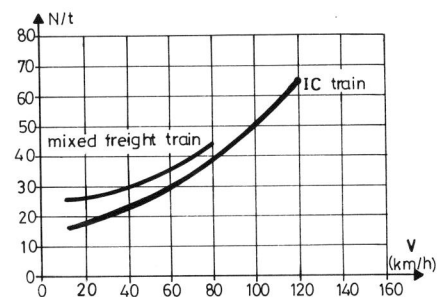

Fig. 12.5. Specific rolling resistance for broad gauge railways

Fig. 12.5. Specific rolling resistance for narrow gauge railways

12.2.4. Resistances developed when running in a tunnel

Compared to open air operation, operation in a tunnel has certain peculiarities caused by sudden increases in pressure with unfavourable influence on passenger comfort, increased aerodynamic resistance, problems arising when trains cross and, finally, the need to ensure proper ventilation.

12.2.4.1. Pressure problems

When a train enters a tunnel, the front section (the head) of the train compresses the air at the entrance giving rise to an overpressure wave (Fig. 12.7), the amplitude of which increases as the train proceeds, reaching a maximum when the rear section (the tail) of the train enters the tunnel. At this moment an underpressure wave is created by the vacuum left behind the speeding train. The overpressure wave at the train front which propagates at the speed of sound along the tunnel, is reflected by its walls and returns in the form of an underpressure wave. With respect to the underpressure wave generated by the tail of the train inside the tunnel, it undergoes corresponding changes and finally returns in the form of an overpressure wave. When all these waves combine, they give rise to pressure fluctuations progressively diminishing in amplitude as a function of time, (23).

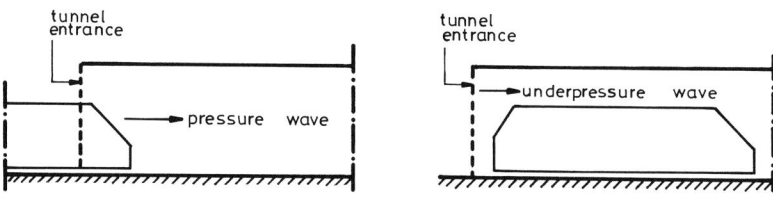

Fig. 12.7. Pressure and underpressure waves when a train enters a tunnel

It should be noted, however, that passenger discomfort is caused not so much by pressure variations as such but by the rate of pressure variation. During abrupt changes of weather the pressure may change by up to 1,300 mm H_2O, and a 1,000 m increase in altitude causes a pressure drop of 1,100 mm H_2O, with no significant passenger discomfort. In contrast, during train motion in a tunnel, pressure changes are much smaller but also much more annoying. The reason lies with the pressure change rate. The human body can adapt to significant changes in pressure, provided that they are not abrupt, (184).

Factors affecting passenger comfort, therefore, include both pressure variation, Δp, and pressure change rate $\Delta p/\Delta t$. Various researches, (184), have shown that passenger comfort is not significantly affected as long as

$$\Delta p \cdot \frac{\Delta p}{\Delta t} < c \tag{12.12}$$

where c is a constant, the exact value of which is different in the various railway networks.

Train Dynamics

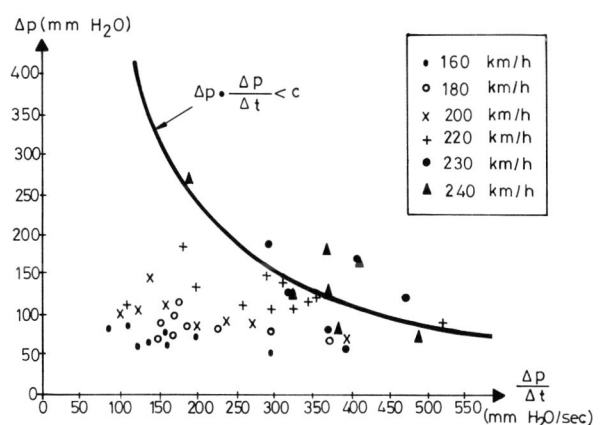

Fig. 12.8. Pressure variation and pressure change rate as a function of speed increase (experimental test results), (184)

Fig. 12.8 illustrates the results of experimental tests in railway tunnels. During the course of these tests it was found that the rolling stock type has an important influence on the values of the pressures developed.

We see that, until a speed of 200 km/h is reached, passenger comfort is not significantly affected. Beyond this value, however, pressure variations and their rates of change become important and currently prohibit very high speeds in tunnels.

12.2.4.2. Increased aerodynamic resistance

In tunnels, aerodynamic resistance is higher. Research by the Swiss and French Railways on type *TEE (Trans Europe Express)* rolling stock have yielded the running resistance as a function of the lateral openings made in the tunnel to reduce aerodynamic resistance, (183):

Tunnel with no openings $R(kg) = 1,107 + 8.25\, V + 0.490\, V^2$

Tunnel with openings
every 250 m $R(kg) = 1,107 + 8.25\, V + 0.224\, V^2$

Tunnel with openings
every 500 m $R(kg) = 1,107 + 8.25\, V + 0.246\, V^2$

Running resistance in
open air $R(kg) = 1,107 + 8.25\, V + 0.158\, V^2$

The above mentioned research has given the total running resistance and the required power for a TEE train weighing 705 t (Table 12.1).

Railway Engineering

Table 12.1.
Running resistance and required power for a TEE train weighing 705 t, (183)

	In the open air	In tunnel — Openings every 250 m	In tunnel — Openings every 500 m	In tunnel — No opening. One train in tunnel
Total running resistance (kg)	6,480	8,170	8,830	14,930
Required power (kW)	2,820	3,550	3,840	6,500

In order to reduce the aerodynamic resistance, efforts are made to reduce the S/Σ_t ratio, where S is the front surface cross-sectional area of the train and Σ_t the effective tunnel cross-sectional area (Fig. 12.9). Thus:

- in single-track tunnels, $\dfrac{S}{\Sigma_t} = 0.30 \div 0.50$
- in double-track tunnels, $\dfrac{S}{\Sigma_t} \sim 0.15$.

It is evident that an excessive reduction in the $\dfrac{S}{\Sigma_t}$ ratio would lead to an inordinate, and expensive, increase of tunnel cross-section, (182).

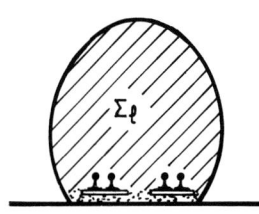

Fig. 12.9. Effective tunnel cross-section Σ_t

A reduction in aerodynamic resistance in tunnels is achieved by reducing the pressure difference between the train front and rear. This has been achieved in the Channel Tunnel, which is composed of two single-track tunnels with communication openings every 375 m (see section 1.6). Calculations have shown that the air passages between the two tunnels will reduce the power required to overcome aerodynamic resistance from 13.5 MW to 5.8 MW at a speed of 140 km/h, (23).

12.2.4.3. Trains crossing in tunnels

When a train crosses another in a tunnel, pressure waves generated by the first strike the other and conversely. As the faster train gives rise to the stronger effects, the slower train is obviously subjected to the heaviest stresses.

Experimental tests by the Italian Railways have shown that aerodynamic effects when two trains cross in a tunnel do not significantly affect passenger comfort, mainly because of their short duration (a few tenths of a second), (184). Human hearing is disturbed by extraneous influences only if they last longer than half a second. With respect to damage of the rolling stock (mainly window glass fracture), the above tests have shown no significant risk at speeds up to 220 km/h, (184a).

12.2.4.4. Tunnel cross-section requirements at high speeds

All the aforementioned reasons entail that the tunnel cross-section increases as speed increases. Table 12.2 gives the effective cross-sectional area Σ_t for various speeds for double-track tunnels and Fig. 12.10 illustrates the dimensions of a tunnel with a running speed of 300 km/h. However, in the design of high speed tunnels (V>250 km/h) emphasis should be paid not only at the distance between tracks (4.50÷4.70 m) and the cross-sectional area Σ_t (80÷100 m²) but also at the performances and mechanical resistances of the rolling stock (particularly the glass parts).

V_{max} (km/h)	160	200	240	300
Σ_t (m²)	40	55	71	~100

Table 12.2. Required tunnel cross-sectional area for a double-track tunnel at various speeds

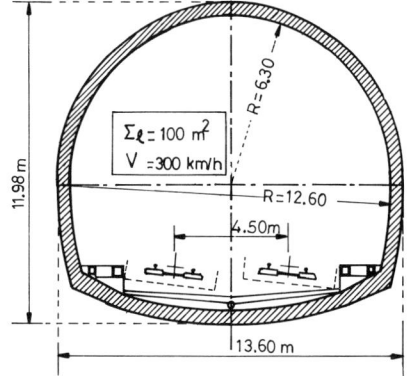

Fig. 12.10. Cross-section of a high speed tunnel

12.2.5. Comparative running resistance between railways and road vehicles

The running resistance of a rail vehicle (passenger or freight) is far lower than that of a road vehicle (Table 12.3). The lower railway running resistance is firstly due to the lower coefficient of rolling friction of the metal wheels on metal rails and secondly to the lower aerodynamic resistance of a train because of its great length.

Table 12.3.
Running resistance for railways, private cars and trucks

	Running resistance (daN) per passenger		$\dfrac{\text{private car}}{\text{rail}}$ ratio
	Private car (4 seats)	Railway (11 vehicles, 820 seats)	5.1
V = 120 km/h	21	4.08	
	Running resistance (daN) per ton transported		$\dfrac{\text{truck}}{\text{rail}}$ ratio
	Truck	Railway	4.3
V = 80 km/h	20	4.6	

12.2.6. Resistance due to track curves

Additional resistance on curves is caused by:
- friction between wheel flange and rail
- wheel slippage on the rails, since the axles of a bogie or of a two-axle rail vehicle are always parallel.

The additional specific resistance r_c occurring along curves can be expressed by the equation

$$r_c \text{ (kg/t)} = \frac{k}{R} \qquad (12.13)$$

where k : parameter with values between $500 \div 1{,}200$, with 800 as the average
R : radius of curvature in the horizontal plane (m).

Train Dynamics

12.2.7. Resistance caused by gravity

In a rail vehicle rolling along a straight level track, the force component perpendicular to the direction of gravity is zero. However, when the plane of the track is inclined (e.g. when the train is running uphill or downhill), a force component R_g develops parallel to the plane of the track (Fig. 12.11), and in the case of an uphill gradient this component is an additional resistance to vehicle motion.

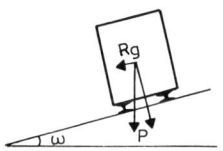

Fig. 12.11. Gravity resistance

As the longitudinal gradient of railway tracks is small and seldom exceeds 20‰, the angle ω is very small and therefore it can be assumed that sin ω = tan ω. Therefore:

$$R_g = P \cdot \sin \omega = P \cdot \tan \omega = P \cdot i \qquad (12.14)$$

where i : the longitudinal gradient.

Resistances due to layout curves and to gravity are commonly unified in a common term called adjusted gradient and usually expressed in kg/t.

12.2.8. Inertial (acceleration) resistance

Resistance forces arising from the acceleration of a train are given by the classical equation of dynamics and depend on the geometrical characteristics and the material of which the vehicles are made.

If α is the acceleration imparted by the traction engine, then the specific inertial resistance r_{in} will be:

$$r_{in} \text{ (kg/t)} = \frac{\alpha}{g} \, q \qquad (12.15)$$

where
q : a mass coefficient taking into account both the fixed and the rotating masses of a rolling stock (shafts, electric motors, transmissions). If M_{rot} are the rotating masses and M the total train mass, then:

$$q = 1 + \frac{M_{rot}}{M} \qquad (12.15a)$$

Measurements have shown that an acceleration of 1 cm/sec² results in an additional resistance force of 1 kg/t, which is approximately as much as that from an uphill gradient of 1‰.

12.3. Specific traction force or starting force of a train

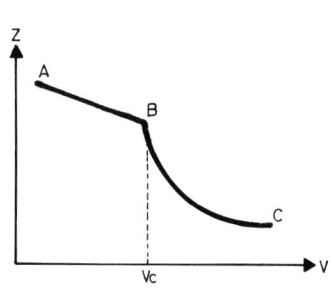

Fig. 12.12a. **Specific traction force - running speed diagram of a diesel train**

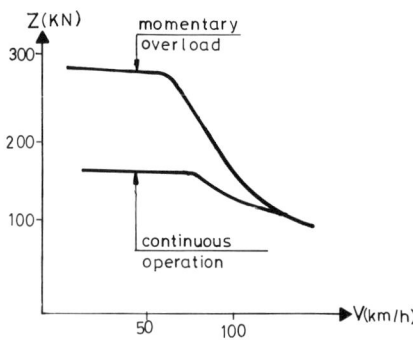

Fig. 12.12b. **Specific traction force - running speed diagram of an electric train**

Specific traction force or starting force Z is the force required to put a train into motion. The starting force should overcome the sum of all resistances, generated during train motion. If all vehicles of a train departed simultaneously, the force Z would have to be very high. In practice, however, this is never the case, because the train does not start as a block, the gaps and couplings between the vehicles (see section 12.9.5.3 below) introducing the elasticity required to start the vehicles in succession.

In diesel traction (Fig. 12.12a), the force developed by the traction engine decreases with increasing speed, while the maximum traction force Z is developed when starting. As speed increases, the traction force decreases, at first linearly (segment AB) and as speed increases further, resistance plummets (segment BC) leveling off at a minimum corresponding to the maximum speed of the pulling vehicle.

Electric trains (Fig. 12.12b) can sustain momentary overloads, in which case the tractive force is greater than that in continuous operation and therefore higher speeds are attainable.

Figure 12.13 illustrates the diagrams of the specific traction force Z as a function of uphill gradient in the cases of passenger and freight trains. Usual values of the specific traction force are $10 \div 20$ kg/t for passenger trains and $10 \div 30$ kg/t for freight trains.

Train Dynamics

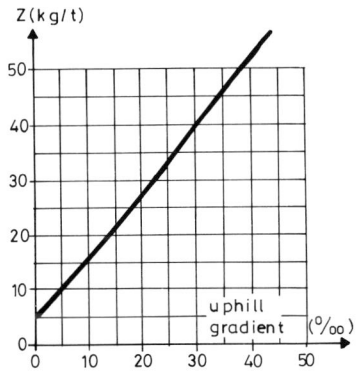
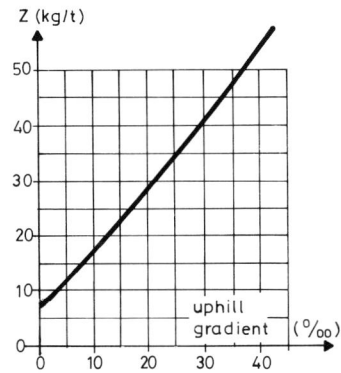

a. Passenger trains

b. Freight trains

Fig. 12.13. Specific traction force Z of a train in relation to uphill gradient

12.4. Adhesion forces

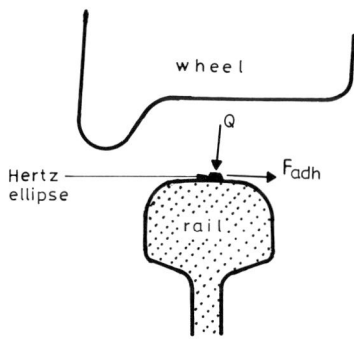

Fig. 12.14. The adhesion force F_{adh}

A fundamental characteristic of rail transport is the metal (wheel) to metal (rail) contact, occurring along an elliptical surface known as the Hertz ellipse (Fig. 12.14, see also sections 2.7 and 5.4.1). Along the Hertz elliptic surface, adhesion forces F_{adh} appear, which are necessary to ensure continuous rotation of the wheel. This requires that the adhesion force F_{adh} be equal to or greater than the traction force Z (Fig. 12.15).

The adhesion coefficient μ is defined as the ratio of the horizontal adhesion force F_{adh} to the vertical wheel load Q :

$$\mu = \frac{F_{adh}}{Q} \qquad (12.16)$$

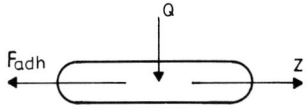

Fig. 12.15. Traction force Z and adhesion force F_{adh}

Railway Engineering

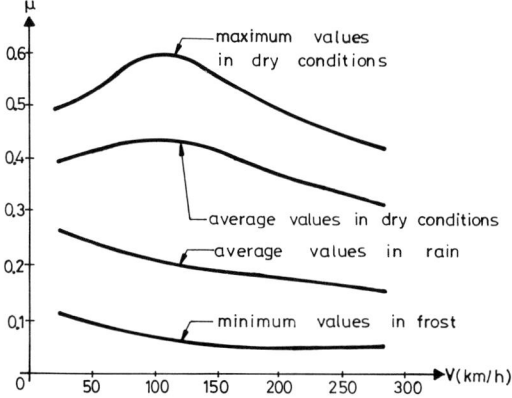

Fig. 12.16. The adhesion coefficient μ in relation to train speed V and climatic conditions

The adhesion coefficient μ depends mainly on climatic conditions but also on train speed (Fig. 12.16). To satisfy the condition $F_{adh} \geq Z$, the required values of μ have been surveyed and are given in Table 12.4, (185), (186).

Table 12.4.
Required values of the adhesion coefficient μ

Traction mode		Braking mode	
V (km/h)	μ_{min}	V (km/h)	μ_{usual}
160	0.3		
200	0.1	0÷200	0.095
300	0.07	200÷300	0.06

Concerning the influence of the various track and rolling stock parameters on the adhesion coefficient, it was found that, (187):
- increasing wheel diameter from 700 mm to 920 mm, caused little increase of adhesion
- changing the transverse slope of the rails on the sleepers (see section 2.9, Fig. 2.13) from 1/40 to 1/20, decreased adhesion by 17%
- increasing per-wheel load from 8 t to 12 t, decreased adhesion by 12%.

A median value of μ as a relation of speed can be obtained from the formula, (182):

$$\mu = \frac{7.5}{V(m/sec) + 44} + 0.161 \qquad (12.16a)$$

Train Dynamics

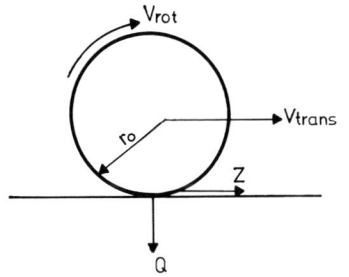

Fig. 12.17. Speeds and forces on a wheel

Finally, for a motor wheel to perform properly, the theoretical peripheral speed of the wheel (Fig. 12.17)

$$V_{rot} = 2\pi r_o n \qquad (12.17)$$

where r_o : rolling radius
       n : number of revolutions

should be greater than the actual translational speed ($V_{rot} > V_{trans}$). Otherwise, there will be:

* braking if $V_{rot} < V_{trans}$
* wheel skid if $V_{trans} < V_{rot} \neq 0$
* wheel lock if $V_{rot} = 0 > V_{trans} \neq 0$

It is evident that constant or increasing train speed requires that the tractive force be equal to or greater than the total resistance developed to train motion.

12.5. Required train power

The tractive force necessary for train motion is ensured by adequate engine power. Engine power is distinguished in nominal power and effective power. *Nominal* power is the power as specified by the engine manufacturer. Part of the nominal power is absorbed by auxiliary devices of the engine and another part is lost during transmission from the motor shaft to the wheels. The remaining power is termed *effective* power and is the part actually available to power the motor wheels and the train as a whole.

Power is measured in either horsepower (hp, ps, cv) or kilowatts (kW). Engine power in horsepower is derived from the relation:

$$N = \frac{Z \text{ (kg/t)} \cdot V \text{ (km/h)} \cdot P(t)}{270} \qquad (12.18)$$

where Z is the specific traction force and P the train weight.

Therefore, it is evident that train power depends on speed, which should be specified every time that a power value is indicated. Table 12.5 gives the power required for operation of various train categories.

Railway Engineering

Table 12.5.
Power required by various types of trains, (181)

Type of train	Weight (t)	Speed (km/h)	Power (kW) at a 5 ‰ gradient
Passenger	800	160	4,400
Freight	1,800	100	6,350
Suburban	190	140	1,050
High-speed (TGV)	418	300	6,850

Power is often indicated per rolling stock unit weight, in which case it is termed *specific power* N_e (kW/t or Ps/t) (see also chapter 13, section 13.6.3). A parameter to determine the course of a train is the distance required to attain a final speed (Fig. 12.18)

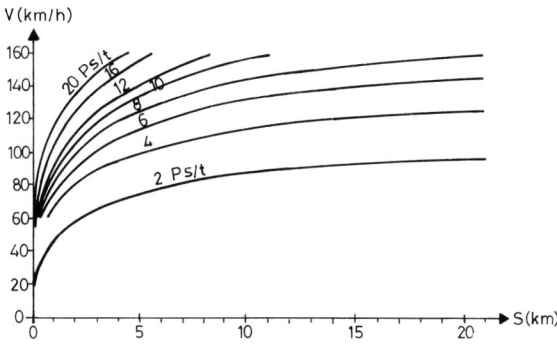

Fig. 12.18. Required distance S as a function of specific power to enable train speed starting from zero to reach a final value

12.6. Train acceleration and deceleration

The values of acceleration and deceleration of a train depend on the type of transportation (passenger, freight) as well as on the distance within which the train must attain its maximum speed. The shorter this distance, the higher the values of acceleration and deceleration, as is the case with the metropolitan and suburban railways. For reasons pertaining to human physiology, maximum acceleration should not exceed 1.2 m/sec^2.

Common *acceleration* values for various types of rolling stock are:

- freight trains: $0.2 \div 0.4$ m/sec^2
- intercity trains: $0.4 \div 0.6$ m/sec^2
- suburban trains: $0.6 \div 0.8$ m/sec^2
- metros: $0.8 \div 1.0$ m/sec^2

Common *deceleration* values for various types of rolling stock are:
- conventional freight trains: 0.10 m/sec²
- express freight trains: 0.25 m/sec²
- passenger trains: 0.40 ÷ 0.50 m/sec²
- suburban railways, metros: 0.60 m/sec²

A critical parameter for passenger comfort is the acceleration variation per unit time, also known as *jerk*. Jerk should not exceed the value of 1.5 m/sec²/sec.

12.7. Train braking

12.7.1. Braking systems

Two braking systems are in use for braking rail vehicles, (183), (187):
- *Shoe (or block)* brakes. They operate with the help of the friction developed on the wheels by the pressure of metal shoes. Both wheels of the axle being braked are provided with braking shoes.
- *Disc* brakes. The braking action is achieved by friction on steel discs or cast iron fixed to the axle. A basic disadvantage of disc brakes is the generation of high temperatures reaching 500°C, (192).

The following methods are used for transmission of the braking force:
- *Air braking*, using changes of air pressure in special conduits, initiated in the driver's cab by operating a valve. This system has the disadvantage that braking is not simultaneous on all train vehicles.
- *Electropneumatic braking*, developed in the 1960s, to reduce the transmission delay of the braking operation to the vehicles in a train. In this system, air pressure is modified simultaneously at all wheels by electrically-actuated air valves at each brake. The system is operated by an electric signal transmitted on a line along the train.
- *Electromagnetic braking*, developed in recent decades so as to confront the great increase in train speeds. In this type, the braking action is applied directly to the rails. Braking is achieved by special shoes with electromagnets which carry a current during braking. Electromagnetic braking may function independently or in combination with other systems.
- *Electrodynamic braking*, doing away with brake shoe wear because deceleration is obtained by converting the electric traction motors into electric generators. The power generated by braking is used for auxiliary purposes. In the case of electric locomotives, the energy recovered may be returned to the power network through the pantograph. The recovered energy is 3 ÷ 6% in intercity trains, 20% in mass transit and freight trains and 40% in trains on high gradient tracks, (182).

Finally, rail vehicles are provided with *anti-skid* devices which monitor wheel rotation and modify the braking force whenever wheel locking is detected. In consideration of train braking, special concern needs to be given to poor adhesion conditions that can be created under certain weather conditions due to rain, ice and leaf deposits.

Finite element analysis gives the possibility to study both the mechanical and thermomechanical behaviour of braking systems (Fig. 12.19).

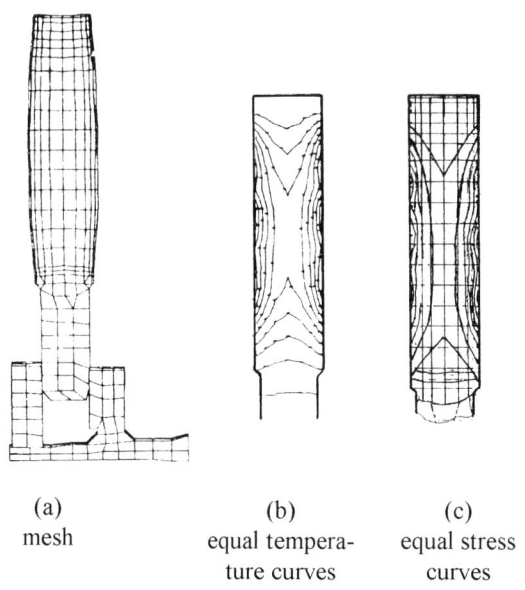

(a) mesh (b) equal temperature curves (c) equal stress curves

Fig. 12.19. Analysis of a disc brake with the use of the finite element method

12.7.2. Braking distance

Empirical formulas have been developed giving the braking distance L for the various train categories, (183), (203).

Freight trains (V < 70 km/h)
This is the oldest equation (Maison's formula):

$$L(m) = \frac{4.24 \ V^2 \ (km/h)}{1{,}000 \ \varphi \cdot \lambda + 0.0006 \ V^2 + 3 - i} \qquad (12.19)$$

where i : track gradient (‰ or equivalent in mm/m). Track gradient is regarded positive downhill and negative uphill.
   φ : friction coefficient depending on gradient. Values are:
      $\varphi = 0.10$, for i < 15‰
         $= 0.10 - 0.00133 \ (i - 15)$, for i > 15‰
   λ : braking percentage, defined as the ratio of the braking force to total vehicle weight and expressing the braking force required for braking 1 t.

Braking percentage λ is a critical factor for the braking distance. Table 12.6 gives values of λ for various types of rolling stock and brakes.

Table 12.6.
Breaking percentage λ for various types of rolling stock and brakes

	λ
Normal brake	
- Tractive vehicles with per axle load P=15÷20 t	80÷95%
- Tracted vehicles with per axle load P=5÷20 t	65÷90%
Emergency brake	
- Tractive vehicles	160÷220%
- Tracted vehicles	130÷220%

Passenger trains (V = 70 to 140 km/h)

The braking distance is given by the Pedeluck empirical formula:

$$L(m) = \frac{\varphi \cdot V^2}{1.09375 \lambda + 0.127 - 0.235 \, i \cdot \varphi} \quad (12.20)$$

with the various parameters defined as in the previous relation.

Diesel - electric passenger vehicles

The braking distance is given by the formula:

$$L(m) = \frac{0.0386 \, V^2}{\gamma - \frac{i}{100}} \quad (12.21)$$

where γ is the deceleration (m/sec²).

Other empirical formulas

Previous formulas, developed by the French Railways, are also in use by the UIC, (186a). However, to overcome the restrictions of formulas (12.19)-(12.21), the German Railways have developed the so called Minden formula for the braking distance, which is:

braking of passenger trains
$$L(m) = \frac{3.85 \, V^2 \, (km/h)}{[6.1 \, \psi \cdot (1 + \lambda/10)] + i} \quad (12.22a)$$

braking of freight trains
$$L(m) = \frac{3.85 \, V^2 \, (km/h)}{[5.1 \, \psi \cdot \sqrt{\lambda} - 5] + i} \quad (12.22b)$$

with the parameter ψ taking values between 0.5 ÷ 1.25 (in relation to the brake type characteristics) which are given by nomographs, (186b).

Another empirical formula, in use by the Belgian Railways, is

$$L(m) = \frac{4.24 \, V^2 \, (km/h)}{\lambda \, \frac{57.5 \, V}{V - 20} + 0.05 \, V - i} \quad (12.22c)$$

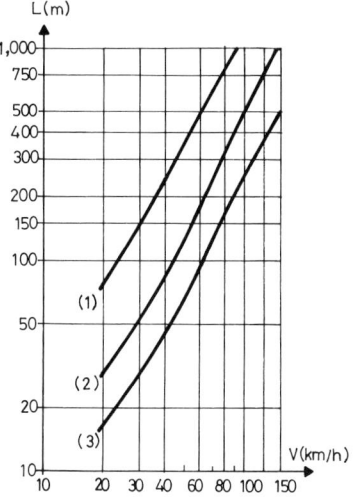

(1) freight train, λ = 40%
(2) rail vehicles, disc brakes, λ = 138%
(3) TEE, disc brakes, λ = 239%

Fig. 12.20. Braking distance in relation to speed (at zero gradient) for medium and low speeds

Figure 12.20 illustrates the braking distance for low and medium speeds and various rolling stock types.

The braking distance derived from the above equations is augmented by at least 10% as a safety margin (depending also on the signalling system). The greater the speed, the longer the braking distance (Table 12.7). Figure 12.21 illustrates the braking distance at high speeds.

Table 12.7.
Braking distance increase in relation to the increasing speed

V (km/h)	Braking distance (m)
160	1,300 ÷ 1,400
200	2,500 ÷ 3,000
260	6,000 ÷ 8,000

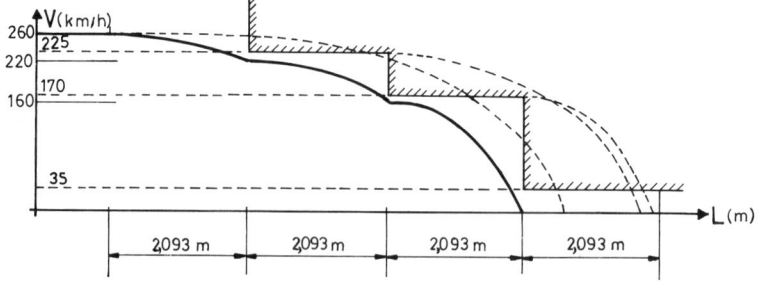

Fig. 12.21. Required braking distance at high speeds (with zero gradient)

234

Train Dynamics

12.8. Line scheduling

The relations developed in this chapter enable the planning of a train schedule. It is first necessary to determine the values of the following basic parameters:
- approved maximum loads
- planned stop locations
- running resistances and adjusted slope
- inertia coefficients of rotating masses
- speed limits due to the track
- speed limits due to the rolling stock
- acceleration on starting
- deceleration on braking.

In addition to the above, operational and commercial parameters should be also taken into account:
- required travel times
- train crossing
- best use of rolling stock

Optimization of line capacity requires the grouping of trains in two categories: fast (passenger) and slow (freight). Within each category, spacing of trains is related to the specific speed and to distances of braking; as previously analyzed, the higher the speed, the greater the braking distance.

Many computer programs have been developed and are in use by the railway services for the accurate calculation of line scheduling. Figure 12.22 illustrates such a scheduling for a double track, (3).

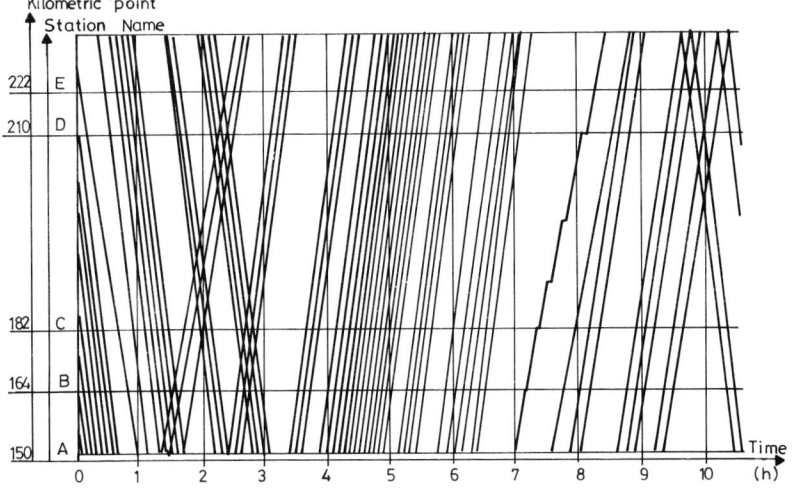

Fig. 12.22. Extract of a line scheduling on a double track

235

Railway Engineering

12.9. Components of a pulled rail vehicle

As already mentioned in section 12.1 of this chapter, every pulled passenger or freight rail vehicle consists of the body, which requires a set of parts and devices for its movement: wheels, axles, bogies, springs, couplers and buffers.

12.9.1. Wheels

On standard-gauge lines, wheel diameter of the pulled rolling stock ranges between $0.85 \div 1.05$ m, with 1.00 m as the average value. On metric lines, wheel diameter averages 0.75 m. As the continuing tendency is to increase per-wheel load, one would expect wheel diameter to increase also. This, however, is not feasible beyond current wheel diameters, because larger wheels would on the one hand increase weight and therefore manufacturing and operating costs, and on the other hand result in a greater vehicle floor height from track level. This would be detrimental to both the stability and the space available in the vehicle (the rolling stock gauge is fixed and cannot be changed), (189).

The wheel diameter of the locomotive is similar to that of the pulled stock. Figure 12.23 illustrates geometrical characteristics of a wheel type.

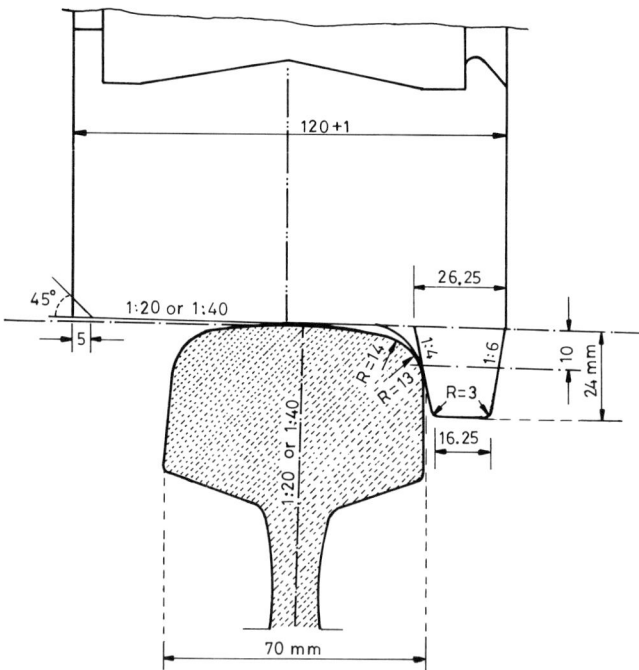

Fig. 12.23. Detail of a rail vehicle wheel

Two main parts can be distinguished in a wheel (Photo 12.1):

- *The tyre*, which is the external part of the wheel and comes in contact with the rail. Since it is subject to high wear, the tyre is made of material highly resistant to wear.
- *The internal disc* of the wheel. The external part of the disc inside the tyre is the *wheel rim*.

Tyre thickness ranges between 65÷70 mm and the tyre is considered worn when wear reduces the thickness to 30 mm.

The first tyre material was soft iron, but it wore out quickly and was difficult to weld properly. Accordingly, it was replaced by hard steel, which, however, should have low brittleness, (188).

Photo 12.1. Tyre and wheel rim of a rail vehicle, (182)

12.9.2. Axles

Wheels are connected in pairs on axles, for which each vehicle includes at least two. The increases in vehicle weight combined with the need to keep subgrade loading within reasonable limits have led to the addition of a third and then of a fourth axle. Four-axle rail vehicles are currently the rule.

The distance δ between the two most distant fixed axles of a vehicle is termed the wheel-base δ of the vehicle (Fig. 12.24). The greater the wheel-base of a vehicle, the more stable it is on straight track but the more difficult it will be to run on curved track. The maximum wheel-base length δ enabling a rail vehicle to operate on a curve of radius R is given by the relation

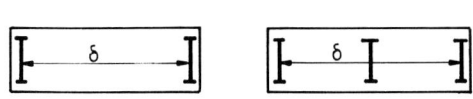

Fig. 12.24. Wheel-base of a vehicle

$$\delta_{max} = 0.30 \sqrt{R}$$

Railway Engineering

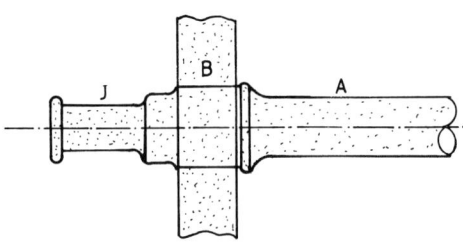

Fig. 12.25. Parts of an axle

An axle is composed of the following parts (Fig. 12.25):
* The axle-journal J which is supported by the bearing
* The wedging region B, which is the part of the axle wedged into the wheel body
* The main body A of the axle, located between the two wheels.

The vehicle load is applied to the bearings and thence transmitted to the journals and the wheels. In order to reduce friction between the journals and bushings, they are lubricated by grease-boxes. The bearings are mounted inside the grease-boxes which also support the vehicle body through suspension springs. There are two kinds of bearings, journal bearings and rolling-contact bearings, which are in turn distinguishable into ball bearings and roller bearings, (195).

Motor axle loading stresses are both torsional and bending, while stresses in non-motor axles are only bending.

12.9.3. Bogies

The increase in the number of axles of railway vehicles gave rise to the need to separate the axles into groups. This is achieved by the bogies, where two or more axles are mounted on the same frame (Fig. 12.26). In commonly used bogies (Fig. 12.27) the axle body and the wheels are rigidly joined, and as a result they rotate at the same angular speed. The bogie frames are connected to the rail vehicle body and to the axles through springs and shock absorbers, providing the vehicle with two suspension directions (Figs. 12.27, 12.28 and Photo 12.2).

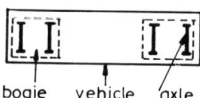

Fig. 12.26. Bogie

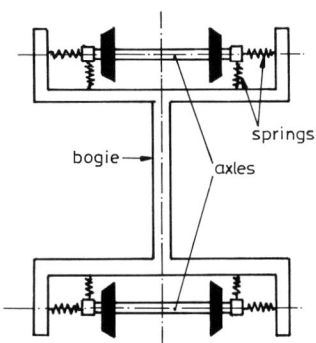

Fig. 12.27. Configuration of conventional-type bogies and location of springs

Train Dynamics

This conventional type of bogie ensures very satisfactory passenger comfort and safety, having been tested in high-speed trains (V_{max}: 250 ÷ 300 km/h). On small radii of curvature, however, tests have shown both wheel slippage and wheel flange contact with the external rail. This has brought about the design and development of new bogie types which include: bogies with self-guided axles, bogies with skid-control axles and bogies with independently rotating wheels, (190), (191).

Photo 12.2. Bogie and springs of a rail vehicle, (182)

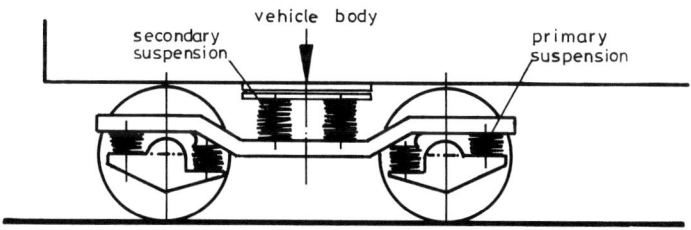

Fig. 12.28. Primary and secondary suspension of a rail vehicle

12.9.4. Springs

Springs are used between parts of the same rail vehicle (Fig. 12.29) as well as between successive vehicles. Depending on their purpose, springs may be: suspension, compression, coupling.

Railway Engineering

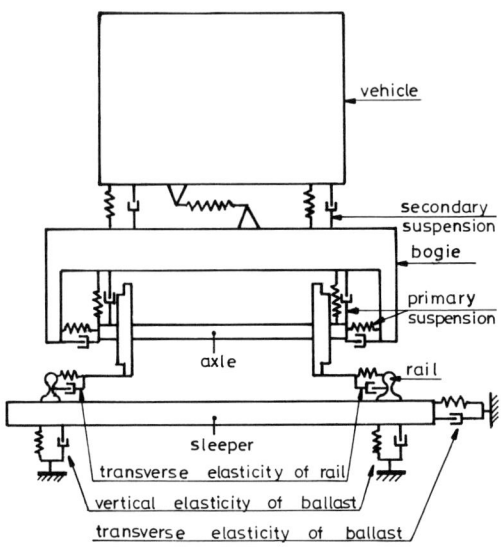

Fig. 12.29. Vertical and transversal elasticities of the rail system

If P is the load applied on a spring and $\Delta\ell$ its length variation, the work energy stored is:

$$W = \frac{1}{2} \Delta\ell \cdot P^2 \tag{12.23}$$

Depending on the purpose of the springs, constraints are set to the maximum value $\Delta\ell$ as follows:
- locomotives, $\Delta\ell$: 10 ÷ 15 mm
- passenger vehicles, $\Delta\ell$: 50 ÷ 150 mm
- freight vehicles, $\Delta\ell$: 15 ÷ 25 mm.

12.9.5. Couplings and buffers

Couplings and buffers are devices used to interconnect rail vehicles to form trains. Their main purpose is to transfer horizontal forces from one vehicle to the other.

12.9.5.1. Springs in couplings

For passenger comfort reasons, springs in passenger vehicle couplings have a low value of $\Delta\ell$: 12 ÷ 20 mm. Contrariwise, in freight vehicle couplings it is $\Delta\ell$: 30 ÷ 50 mm.

12.9.5.2. Springs in buffers

As opposed to coupling springs, buffer springs in passenger vehicles should have a high value of $\Delta\ell$ (ranging between $50 \div 70$ mm), to absorb the various shocks and vibrations thoroughly and quickly. A similar requirement is not necessary for freight vehicles, in which $\Delta\ell$ ranges between $30 \div 50$ mm.

12.9.5.3. Couplings

Hook couplings (Fig. 12.30) were formerly used to connect the vehicles in a train. However, automatic couplings are implemented today, automatically ensuring the coupling of successive rail vehicles, in particular the connection of brake air pipes and electrical circuits.

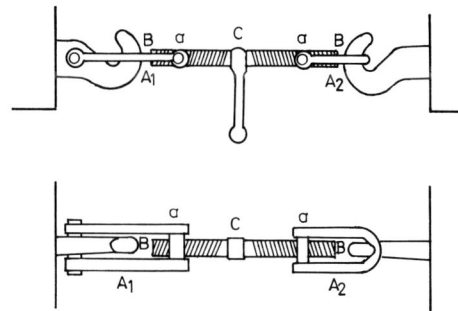

Fig. 12.30. Hook coupling

12.9.5.4. Buffers

Buffers (Fig. 12.31) are employed to keep constant spacing between rail vehicles and to absorb shocks. In standard-gauge lines, buffer height above rail level is $0.90 \div 1.25$ m.

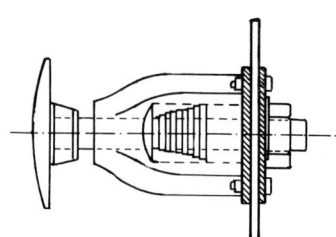

Fig. 12.31. Buffer detail

12.10. Dynamic stability of a rail vehicle

The dynamic stability of a rail vehicle depends on the motions occurring in the transverse direction Y of the vehicle (Fig. 12.32). Bogies have three degrees of freedom (Fig. 12.33), to which should be added the two degrees of freedom of each of the two axles, i.e. transverse motion and rotation around the vertical axis (Z). Vehicle motion therefore shows a total of seven degrees of freedom.

Railway Engineering

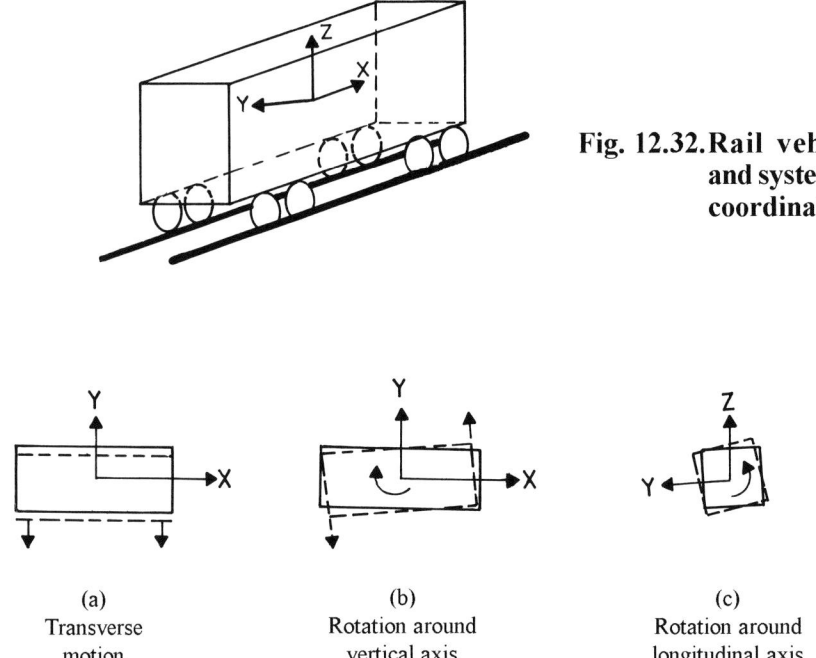

Fig. 12.32. Rail vehicle and system of coordinates

(a) Transverse motion

(b) Rotation around vertical axis

(c) Rotation around longitudinal axis

Fig. 12.33. Degrees of freedom of a bogie

The vehicle instability conditions are found by solving the equation of dynamics:

$$[M][\ddot{x}] + [C][\dot{x}] + [R][x] = 0 \qquad (12.24)$$

where [M] : the mass matrix
       [C] : the viscosity (damping) matrix
       [R] : the stiffness matrix
       [x] : the column matrix of the degrees of freedom, i.e.

$$[x] = \begin{bmatrix} x_1 \\ x_2 \\ x_3 \\ x_4 \\ x_5 \\ x_6 \\ x_7 \end{bmatrix} \qquad (12.25)$$

Solving equation (12.24) leads to an auxiliary equation of the 14th degree. Let λ_i be the real roots of the auxiliary equation. The critical speed is the one corresponding to a change of sign of the product $\Pi\lambda_i$, (108).

The above analysis was based on the assumption that all phenomena are linear. However, this is not exactly the case, particularly as the critical speed is approached and conditions of instability are created. The above theoretical analysis should be confirmed by experimental tests.

Combining the results of the theoretical and experimental analysis, (181), (182), (190), it was found that the critical speed increases as:
- bogie axle spacing increases
- the moment of inertia of the bogie around the vertical axis decreases
- bogie mass decreases.

Since the last two requirements are in conflict with the first, a compromise is sought during rolling stock design. Bogie mass, of course, can only be reduced to a certain limit, so that vehicle loads are adequately supported.

Both theoretical analysis and experimental tests have shown that the critical speed of a bogie with new wheels is higher than that for a bogie with worn wheels. Indeed, critical speed decreases rapidly beyond a certain degree of wear. Wheels are accordingly subjected to limited rectification on a lathe after a certain kilometric distance (normally once a year).

The critical speed for passenger rolling stock in current use has been found to be well over 300 km/h, a fact conferring to rail vehicles high dynamic stability even at the highest speeds.

12.11. Vehicles with body tilting on curves

As explained in the chapter on track layout (section 9.2.2), the coexistence of passenger (fast) and freight (slow) trains along a curve must be ensured. The normal cant h implemented does not fully compensate centrifugal forces of fast trains and a cant deficiency h_d exists, which is the difference between the theoretical value of cant for maximum speed and the normal cant:

$$h_d = h_{th}(V_{max}) - h$$

with serious consequences on passenger comfort.

To deal with the problem, rail vehicles have been manufactured recently with bodies capable of tilting relative to the wheel-base (Fig. 12.34c) when running on curves. Therefore, concerning the passenger level a cant is applied very close to the theoretical cant h_{th} and clearly exceeding normal cant of the track layout. This tilting body feature drastically reduces (and often eliminates) cant deficiency.

Railway Engineering

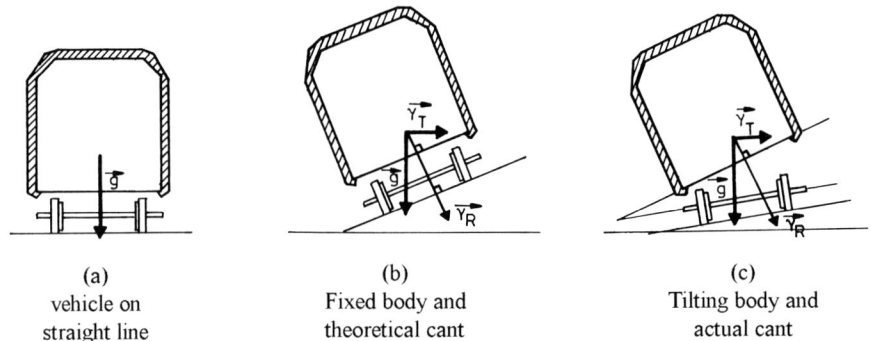

(a) vehicle on straight line

(b) Fixed body and theoretical cant

(c) Tilting body and actual cant

Fig. 12.34. Positions of a rail vehicle and accelerations developed

Two tilt-body rail vehicle technologies have been developed:
- The *passive* method, whereby the vehicle suspension rises when running on curves so that the turning point of the vehicle remains above its center of mass. This method has been applied to the Spanish Talgo and permits angles of $3° \div 5°$ between body and axle.
- The *active* method, in which the turning angle between body and axle is greater $(8° \div 10°)$ and is set as a function of the unbalanced centrifugal acceleration. As the train enters the transition curve from the straight line, the accelerations developed at the bogie are sensed by accelerometers. Instructions to begin rotation of the body around its axis are issued by an electronic device located at the head of the train. The turning angle is a function of the unbalanced centrifugal acceleration. This technique has been applied to the Italian Pendolino, the Swedish X2000, the German VT610, and others, (193).

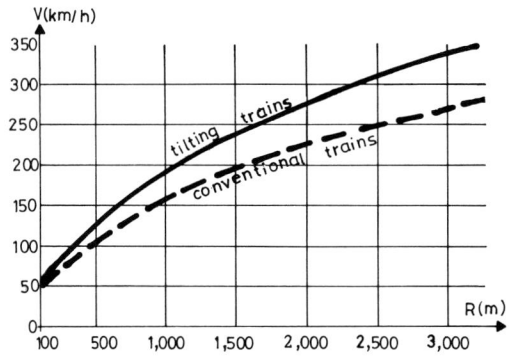

Fig. 12.35. Running speed in relation to the radius of curvature for conventional and tilting trains, (182)

Figure 12.35 illustrates the running spead in relation to the radius of curvature for conventional and tilting trains. It can be seen that in existing layouts, tilting trains can have speeds greater by $20 \div 30\%$ in comparison to conventional trains. However, special consideration in the case of tilting trains must be given to the dynamic envelope in which the vehicles run and the structure gauge of tunnels to be crossed.

13 Diesel and Electric Traction

13.1. The various traction systems

As mentioned in the preceding chapter, the purpose of the pulling vehicle is to pull the train. When a rail vehicle is purely a train pulling engine, it is known as a locomotive. Traction power generation may rely on steam, diesel, or electric motors.

The first power generation medium used for traction was steam. Indeed, the spread of the railways was primarily due to the industrial revolution of steam. The first steam vehicle appeared in 1804 and was used for railway traction in the 1830s. For more than 120 years the steam engine was the principal means for pulling trains.

13.2. Steam traction

13.2.1. Operating principle of the steam engine

Steam engine operation is based on the following principle (Fig. 13.1): The wheel T is connected to the rod MD by the crank TM. The rod MD is con-

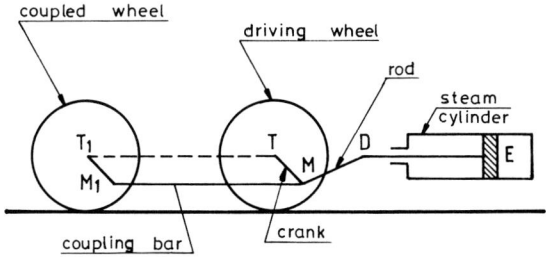

Fig. 13.1. Operational principle of the steam engine

nected to the piston rod DE of the steam cylinder, thereby converting the reciprocating motion of the rod DE generated by steam power into wheel rotation. The wheels of the motor axle, connected to the rods (one on each side) are known as the main driving wheels. A single driving axle is not sufficient to provide the required traction force and therefore other driving axles are also provided. Since the latter, however, cannot be directly connected to the rod, their wheels are coupled to the main driving wheels by rods called coupling rods which, in turn, are successively connected to the wheel cranks TM, T_1M_1, etc. These axles are known as coupled axles. The coupled axles and the axle of the main driving wheels are the driving axles. The total number of coupled axles seldom exceeds five or at most six, but is never less than two.

A steam locomotive may use coal or petroleum as fuel. According to the aforementioned operating principle, the thermal energy liberated by either coal or petroleum is stored as steam pressure dynamic energy and is converted, when necessary, to train kinetic energy.

13.2.2. Main parts of a steam locomotive

The main parts of a steam locomotive are:
- The vehicle, mainly comprising the rolling devices, the frame, the coupling devices, buffers, suspension, etc., as well as the driver's cab, in which all equipment and instruments for locomotive operation and control and for running the train as a whole are located.
- Steam generation equipment, i.e. the boiler and associated parts.
- The engine, i.e. steam cylinders, pistons, slide valves, distribution devices.
- Various auxiliary systems, e.g. air compressors for braking, central heating, sand boxes to increase adhesion between driving wheels and rails, lubrication and braking devices, several safety systems, etc.

13.2.3. Disadvantages and abandonment of the steam locomotive

Presently, steam traction is commercially employed only on certain railway lines of Africa and Asia, whereas in Europe and Northern America it is by now a museum item and the few remaining steam locomotives are used only as a tourist attraction. The reasons that steam locomotives are no longer used are many, (181):
- Low fuel efficiency. Only about 6% of the energy liberated by coal combustion is used for train traction.
- Poor technical performance. Steam locomotives cannot exceed 3,000 horse-power and a maximum operating speed of $120 \div 140$ km/h.
- The need to maintain a large number of water supply facilities.
- High maintenance cost.
- Time-consuming fuel replenishment procedure. A steam locomotive can operate only $12 \div 14$ hours a day.

Diesel and Electric Traction

- Increased fire hazard.
- Harmful to the environment (atmospheric pollution, noise).

13.3. From steam traction to diesel traction and electric traction

13.3.1. From steam traction to diesel traction

Diesel traction of trains was introduced shortly before World War II but was systematically developed after the end of the War. Diesel locomotives are driven by a diesel internal-combustion engine. In comparison to steam traction, diesel traction offers far higher efficiency, a lower operating cost (by almost 50%), much better performance (power, speed), cleaner operation and improved passenger comfort, convenience and less strenuous work for the driver.

13.3.2. From diesel traction to electric traction

The emergence of electric railway traction dates back to 1879. The first uses of electric power in vehicles were restricted to urban service, with the development of the electric tramways between 1890-1914.

Electric traction was first introduced in conventional railways in 1900, when it was adopted in the Paris and London metro lines and in the mountain lines of Switzerland.

Since 1920 electric traction has been extensively used, especially after 1950. Its operating cost is 35% lower than diesel traction, but it necessitates a higher initial expense due to the permanent plant required (see section 13.7.1). Electric traction is accordingly used only on high-traffic lines.

13.3.3. Gas turbine locomotives

Gas turbine locomotives were developed in the 1960s and early 1970s. The basic element of a gas turbine locomotive is the gas turbine engine, operating by the expansion of overheated and compressed gaseous combustion products. After the energy crisis of 1973, gas turbines were no longer cost-effecient due to their high energy consumption, and the instances of their current use are few.

13.4. Diesel traction

13.4.1. Operating principle of the diesel engine

The basic element of the diesel engine (Fig. 13.2) is the cylinder C, inside which is moving the piston P by reciprocating motion. This reciprocating mo-

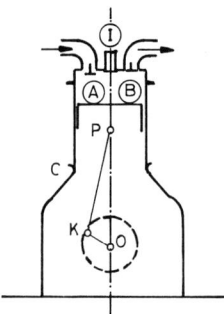

Fig. 13.2. Schematic representation of diesel engine

tion is transmitted as rotary motion by the rod PK and the crank OK to the main driving shaft OP. On the cylinder cover are located the valves A, B and I, performing functions in the following order:

- Suction
- Compression and injection of gaseous fuel
- Combustion and expansion of gaseous combustion products
- Exhaust

A fourth valve is provided near the other three, letting in compressed air to start the engine. All four valves are spring loaded and are opened and closed at appropriate times by levers and a camshaft.

The aforementioned description refers to a single-action engine in which suction, injection and exhaust occur in the cylinder chamber on the same side of the piston, specifically on the upper side.

The motor function of a diesel engine is carried out in four cycles, as follows:

1. *Suction stroke.* This involves the time period in which the piston, starting from the upper end of its travel (UET) with valve A open and valve B closed, descends to the lower end of its travel (LET), while cylinder C is filled with fresh air.

2. *Compression stroke.* This corresponds to the ascent of the piston from the LET, when valve A and valve B close and remain closed, to the UET. Thus, the air pressure in cylinder increases from 1 atmosphere to $30 \div 40$ atmospheres, causing the temperature to reach $400 \div 500°$ C. Shortly before the piston reaches the UET, fuel is injected under pressure into the cylinder. The fuel droplets subsequently ignite at the moment when the piston reaches the UET. The pressure within the cylinder C then reaches $50 \div 80$ atmospheres and the temperature reaches $1,800 \div 3,000°$ C.

3. *Expansion stroke.* This corresponds to closed valves and to a piston travel from UET to LET under the action of expanding gases. Valve B opens at the appropriate moment.

4. *Exhaust stroke.* This covers piston travel from LET to UET with valve B open. Combustion gases are forced out, while a large part of the gases is released from the cylinder during the first moments of this stroke. The remainder is pushed by the piston during its ascent to the UET, when valve A opens.

Upon completing the above cycle, the piston returns to its initial position and the process is repeated. It is possible, however, to complete the entire cycle in two strokes, in which case we have a two-stroke engine. With respect to the number of cylinders, there are diesel motors with 4, 5 or 8 in-line cylinders, or $8 \div 12$ cylinders in a V arrangement. Motor speeds range from low (750 r.p.m.) to high ($1200 \div 1500$ r.p.m.). Low-speed motors are heavier for the same power. In order to withstand high temperatures, cylinders have double walls for water circulation in between.

13.4.2. Transmission systems

In diesel locomotives, drive power transmission from the motor to the wheels is achieved by the following methods:
- By *hydrodynamic* transmission and *hydrodynamic* speed shifting (e.g. of the Voith type).
- By *hydrodynamic* transmission and *mechanical* speed shifting (e.g. of the Mekydro type).
- By *electrical* transmission, in which case the diesel engine drives an electric generator in turn driving series electric motors, which are mechanically joined with their driving wheels. In the case of an electrical transmission, no gear boxes are employed and operating conditions are identical to those in electric locomotives. Diesel locomotives of this type, termed diesel-electric locomotives, essentially are direct current-generator plants supplying the motors of the driving axles. If traction requirements are high, several diesel locomotives in series may be used in the same train.
- By other means, e.g. by *hydrostatic* transmission, or by purely mechanical transmission.

13.4.3. Requirements from a diesel locomotive

A diesel engine should meet the following requirements:
- Pulling capability of medium and heavy loads on a level track, uphill, or downhill, with a high transmission box efficiency at medium and high speeds and at a full or almost full train load.
- Overload capability, on the one hand in the low speed range, and on the other hand uphill at full load.
- Capability to brake with no slippage at high speeds, as well as to keep within speed limits downhill without using mechanical brakes.
- Motor operation within the favourable operating region.
- High reliability and low maintenance cost.

13.4.4. Advantages - disadvantages of diesel traction

Diesel traction in comparison with electric traction offers the following advantages:
- lower track installation cost
- autonomy.

However, diesel traction has the following disadvantages compared to electric traction:
- lower mechanical performance (power, force, speed)
- higher energy consumption
- more air pollution and noise
- higher maintenance expenses.

13.5. Electric traction

In contrast to diesel traction, where the energy required for train operation is generated within the diesel locomotive itself, the energy needed for electric traction is transmitted to the electric locomotive by an external subsystem, the power supply subsystem.

13.5.1. Power supply subsystem

The power supply subsystem includes:
- *substations*, where the voltage is stepped down and (in certain electric traction systems) the alternating current (AC) frequency is converted or the AC is rectified into direct current (DC)
- *transmission lines* or *conductor rails* to convey the electric energy from the substations to the electric locomotive.

The electric substations may obtain electric power:
- either from the national high-voltage power network at a frequency of 50 Hz in Europe or 60 Hz in the USA
- or from a separate high-voltage distribution network, at a frequency (commonly $16\frac{2}{3}$ Hz), considerably lower than that of the national network.

  This separate network may be connected to the public network or may be independent, i.e. have its own power generating plants.

Therefore, when planning the electrification of a railway line (existing or under construction), the proximity of the public power network to the line as well as the energy available from the power network should be considered.

In substations, the characteristics of the electric energy obtained from the power network are changed (voltage reduction and/or frequency conversion and/or rectification from AC to DC), and the converted energy is channelled through the transmission line to the rail vehicles. Substation spacing ranges from $10 \div 70$ km and mainly depends on the electric traction system but also on line traffic load.

As a rule, the transmission line from substations to vehicles is in single-phase configuration. Electric traction engines obtain electric power from a conductor which may be:

- either an *overhead line*, as used in railways and (sometimes) in metros
- or a *conductor rail* (one or two), used in metros and some suburban railways.

When only one transmission line or conductor rail is provided, current return is effected through the rails. Either one or both rails may be employed.

13.5.2. Traction subsystem

The traction subsystem includes the electric traction engine with all its equipment and devices. In this subsystem, electrical energy is converted into mechanical energy, which is used to operate the train.

In the case of an overhead transmission line, electric power is transferred from the line to the vehicle through a *pantograph*. In the case of third or fourth rail conductors, *collector shoes* on the vehicles pick up the power (see section 13.8.5).

13.5.3. Requirements of the two discrete subsystems

The two aforementioned subsystems, power supply and traction, have different requirements and, depending on the priority assigned to energy transmission (power supply subsystem) or energy use (traction subsystem), various electric traction systems have been developed.

13.6. Electric traction systems

13.6.1. Direct current traction

Direct current is superior to alternating current as regards the traction subsystem. For a long time, therefore, (since the beginning of this century until about 1950), priority was given to good motor operation. As series-excited DC motors until recent times offered the best operating conditions for railway traction, railway designers sought an electric traction system using direct current. Direct current, however, cannot be transformed. Early electric transmission systems, therefore, operated at the same voltage as the traction motors. The main voltages employed were:
- *750 V*, mainly for transmission on third and fourth rail systems
- *1,500 V*, more widespread than other voltages
- *3,000 V*.

The above voltages are far lower than those employed on public power networks (150,000 V, 220,000 V and 280,000 V) and too low for efficient power transmission. DC railway traction therefore necessitates large cross-sections of the transmission line (400 ÷ 900 mm^2) and closely-spaced substations. Sub-

Railway Engineering

station spacing is 15÷20 km in the case of 1,500 V and 35÷40 km in the case of 3,000 V.

Therefore, direct current railway traction, though superior as regards the traction subsystem, proves inferior when it comes to the power supply subsystem. DC traction is presently used in about 50% of electric railway lines worldwide and has mainly been used in France, Spain, Italy, Russia, Japan, certain parts of the U.K., etc., (Fig. 13.4, p. 254).

13.6.2. Alternating current traction

Alternating current is superior to direct current as regards the power supply subsystem, but encounters problems in the traction subsystem. AC motors meeting the requirements of traction engines are series-excited AC motors with collector, which, however, face problems proportional to the AC frequency. The need therefore initially arose to use AC at a frequency lower than the 50 Hz used on public power mains.

13.6.2.1. Alternating current traction at 15,000 V, $16\frac{2}{3}$ Hz

In this system, electric substations may obtain power from either of two sources:
* From the national power network (at a frequency of 50 or 60 Hz), in which case there is voltage reduction in the substations and a frequency conversion.
* A separate network carrying low-frequency AC, in which case there is only a voltage reduction in the substations.

AC traction at 15,000 V, $16\frac{2}{3}$ Hz is currently used in Central Europe (Germany, Austria, Switzerland) where substations are supplied from special low-frequency AC power plants, and in Northern Europe (Sweden, Norway) where substations are supplied from the 50 Hz national power network. This AC traction system corresponds to 20% of electric railway lines worldwide (Fig. 13.4).

In 15,000 V, $16\frac{2}{3}$ Hz electric traction, substations are spaced 50÷60 km apart and transmission lines have considerably smaller cross-sections than in DC traction. This system, however, has the disadvantages referred to in connection with motors, mainly their great susceptibility.

13.6.2.2. Alternating current traction at 25,000 V, 50 Hz

To overcome the disadvantages of the two previously described systems, it is necessary to seek a traction system that combines the advantages of both systems without presenting any of their disadvantages. This was achieved after 1950 with the development of efficient and lightweight ignitron rectifiers (later superseded by thyristors, which in their turn have been superseded during the

1980s by GTO technology, see section 13.10.3 below) which could be carried on board. In this system, substations are supplied from the national network and simply step the voltage down to 25,000 V, 50 Hz, which is transmitted to the locomotive through the transmission line. In the locomotive the voltage is again stepped down, rectified, and applied to the series-excited DC traction motors.

The 25,000 V, 50 Hz system represents 30% of electric railway lines worldwide and is almost exclusively used in new electric railway traction facilities. Substations are $60 \div 100$ km apart and transmission conductors have cross-sections $3 \div 5$ times smaller than in DC systems. This system moreover enables the use of series DC motors, which are superior for driving trains.

A comparison of the construction cost for traction systems using 1,500 V DC and systems using 25,000 V, 50 Hz AC, yields figures lower by 30% for the latter than for the former (Fig. 13.3).

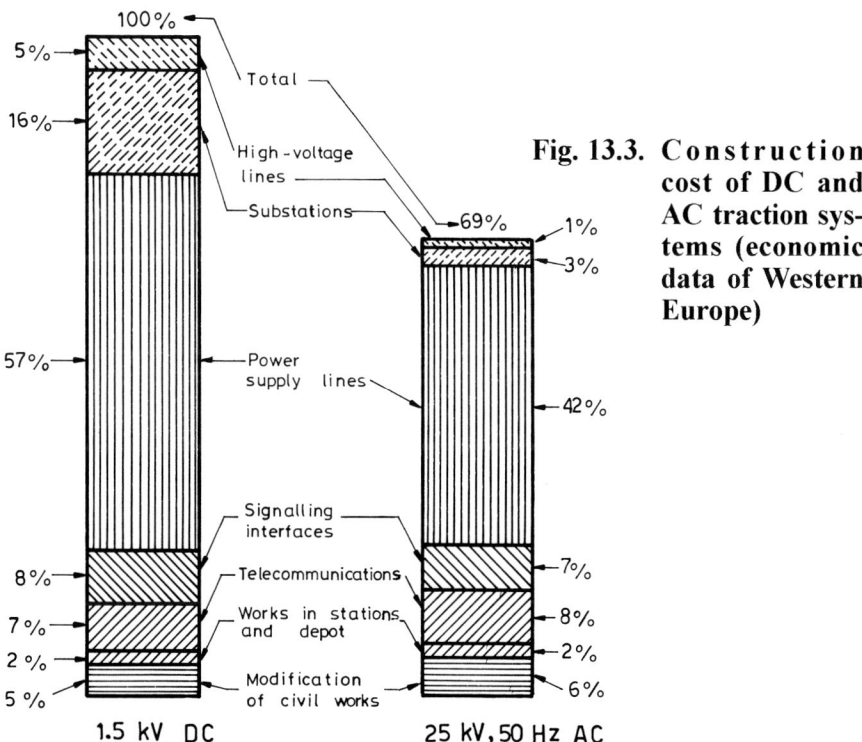

Fig. 13.3. Construction cost of DC and AC traction systems (economic data of Western Europe)

Figure 13.4 illustrates the traction systems for the various European countries and Fig. 13.5 the basic components and characteristics of each system.

Railway Engineering

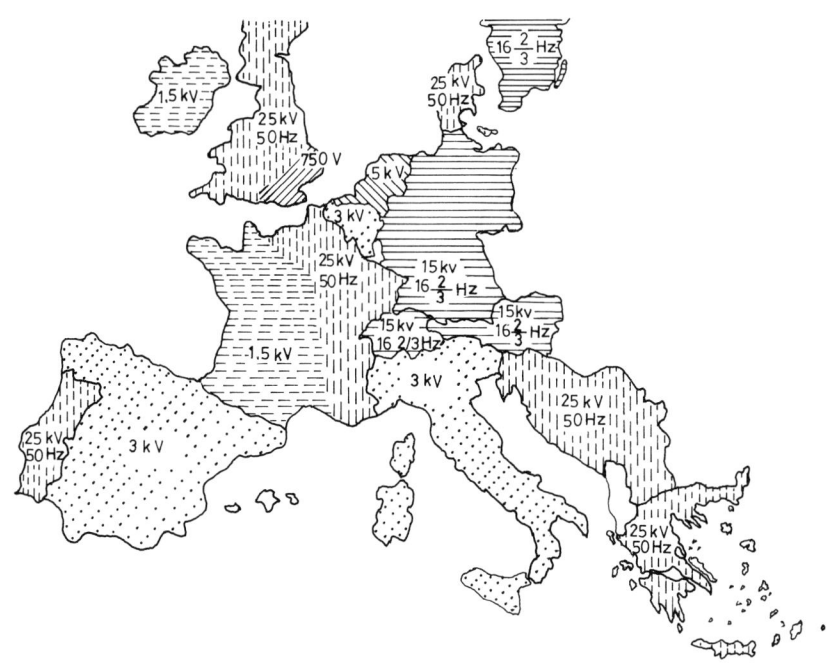

Fig. 13.4. Traction systems for various European countries

13.6.3. Advantages and disadvantages of electric traction compared to diesel

A basic advantage of the electric locomotive is its specific power ($50 \div 55$ kW/t), more than double the specific power of the diesel locomotive ($20 \div 25$ kW/t) (Fig. 13.6).

Electric engines can sustain momentary overloads (when starting, on steep gradients, etc.), whereas diesel engines are not capable as long as acceptable lifetime and maintenance cost constraints are considered.

Furthermore, along lines traversing high-altitude regions, no power drop is observed with electric traction engines. This is not the case with diesel engines, as the air entering the engine is significantly reduced.

In the case of long tunnels, electric traction is mandatory due to the limited air supply.

Finally, electric engines cause little, if any, atmospheric pollution, while maintenance is far simpler and easier than with diesel locomotives. Nevertheless, it should be noted that even diesel trains pollute much less than automobiles (at the ratio of 1 : 14 per passenger-km) (Table 13.1).

Diesel and Electric Traction

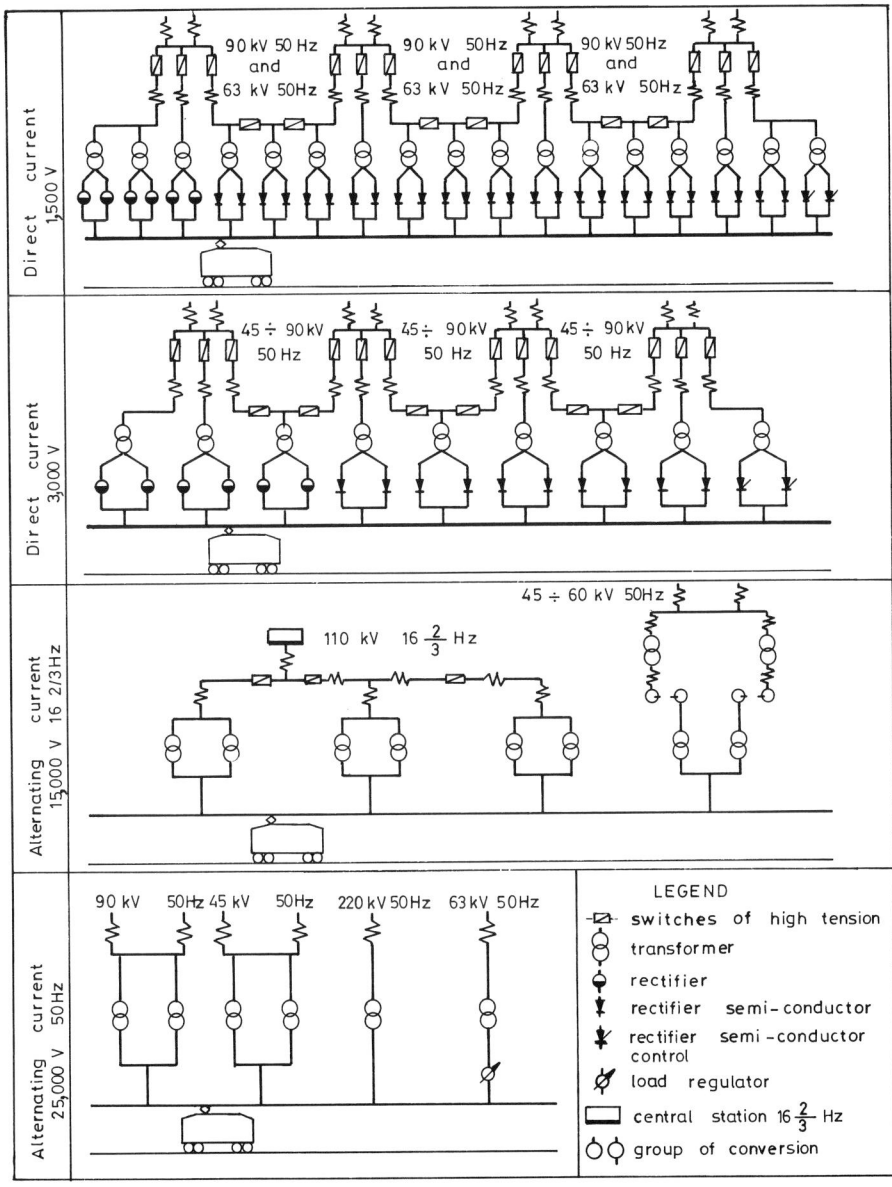

Fig. 13.5. Basic components and characteristics of the various electric traction systems, (6)

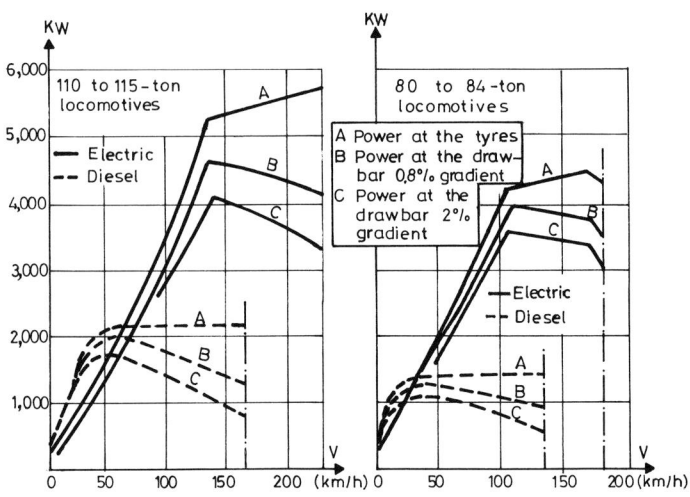

Fig. 13.6. Comparative power of electric and diesel traction engines

**Table 13.1.
Atmospheric pollution caused by the automobile and the diesel train, (23)**

	Number of seats	Equivalent index $(CO + NO_x)$
Automobile	5	8.72
Diesel train	500	0.60

13.7. Feasibility analysis of electric traction

13.7.1. Feasibility analysis parameters and procedure

Two cost factors should be taken into consideration:
- Construction expenses, including transmission lines and substations, which do not depend on traffic
- Operating and maintenance expenses, which increase with traffic.

The quantity commonly studied is the total annual expense as a function of line's traffic, an indice of the latter being the energy expended annually per kilometre of line. Figure 13.7 illustrates a comparative presentation of the annual expenses of diesel and electric traction.

Diesel and Electric Traction

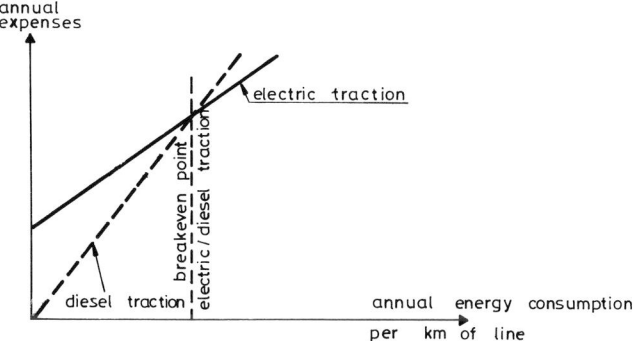

Fig. 13.7. Annual expenses as a function of energy consumption per kilometre of line for diesel and electric traction

We see that at low traffic, electric traction is not cost-effective. However, as the point beyond which electric traction becomes cost-effective is approached, a more detailed investigation of the problem is required.

The period of analysis is usually the following $20 \div 25$ years, and expenses as a whole (initial construction expense and annual operating expense) are converted for each traction system to constant prices by the present value method, (41). Figure 13.8 illustrates the form of the curves referring to each traction system, enabling an accurate determination of the traffic load beyond which electric traction becomes cost-effective (see also chapter 9, section 9.14.1).

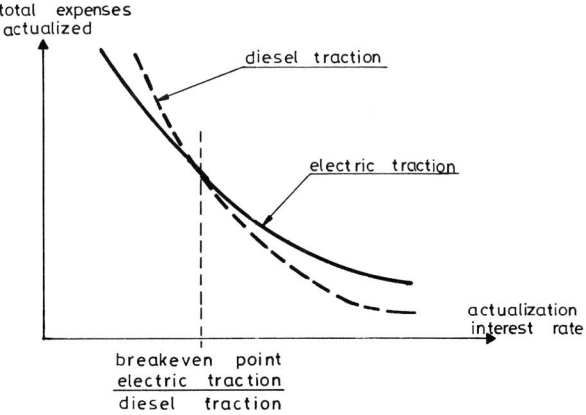

Fig. 13.8. Feasibility analysis for determination of the critical traffic load on a line beyond which electric traction becomes cost-effective

Railway Engineering

A feasibility analysis of electric traction involves many uncertainties, particularly with respect to the price of liquid fuel in the next $20 \div 25$ years, the actualization interest rate whereby the various expenses are converted to constant prices, the length of the feasibility analysis period, etc. It is accordingly advisable to also perform a sensitivity analysis (which aims to examine the impact of the variation of one parameter to the result of the feasibility analysis), (41).

13.7.2. Criterion for selection of the lines to be electrified

The foregoing analysis is not applicable in every case and moreover in most cases the need arises to reach a conclusion easily and quickly as to whether or not electrification of a particular line is advisable. The various railway networks have accordingly adopted simple criteria to this effect, the most widely used being the number of trains on a line or (more precisely) the energy consumption per km of line.

The criteria in question vary from one network to the other, since particularities of each country as regards cost of labour, cost of energy, financial investment cost, etc. are involved.

A criterion, which, however, can only be used to make a first approximative estimate, is the number of trains running on the line. For example, until 1973 (when the cost of energy was low), a line had to be run daily by at least 30 trains per direction to qualify for consideration of electrification cost-effectiveness. After the energy crises of 1973 and 1979, this criterion became around 15 trains per direction daily.

However, given that a train may consist of a small or a large number of vehicles, a criterion was sought which would take into account the number of vehicles. Such a criterion is the energy consumption per km of line. For instance, the French Railways consider in principle an annual consumption of 70,000 kW/km of line as the electrification cost-effectiveness threshold, while the German Railways estimate this limit at 150,000 kW/km, (198), (203). Criteria, therefore, from one network to the other may differ significantly.

When the traffic load or the energy consumption on a particular line exceed the above limits, a detailed feasibility study should be conducted, as described in section 13.7.1, before any decision to electrify the line is made.

13.8. Transmission line

13.8.1. Parts and components of the transmission line

The transmission line concept includes:
- Feeder conductors, contact conductors (touching the pantograph), suspension wire rope, guy wires

Diesel and Electric Traction

Photo 13.1. Pole-supported transmission lines

Photo 13.2. Frame-supported transmission lines

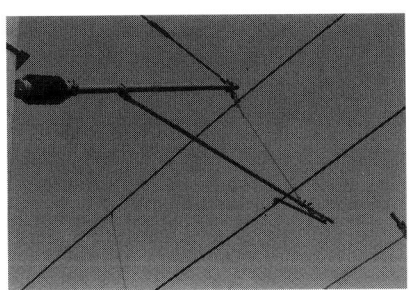

Photo 13.3. Post brackets and insulators

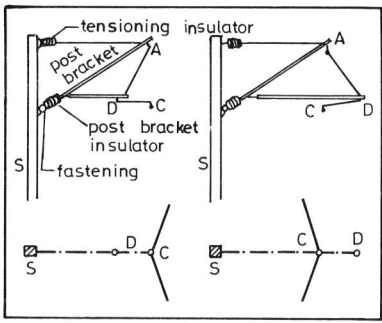

Fig. 13.9. Insulators and post brackets (C is the supply point)

- Conductor support structures, which may consist of poles (Photo 13.1) or frames (Photo 13.2).
- Insulators, post brackets (Photo 13.3, Fig. 13.9), tensioning devices (usually every 1,200 m), counterweights, various mounting hardware, wires connecting the poles to the transmission line and to the ground, conductors for connection to the substations.

As illustrated in Fig. 13.9, the overhead transmission line is suspended from the post brackets which in turn are mounted by insulators on supporting poles, erected $2.20 \div 3.00$ m from the track axis. The post brackets are usually zinc-plated steel pipes.

259

13.8.2. Calculation of the transmission line cross-section

Calculation of the cross-section and other characteristics of the transmission line is performed on the basis of the permissible voltage drop from the substations to the locomotive switchboards, allowing a fluctuation of no more than 10% from nominal value.

Theoretical calculation of the voltage drop is based on the assumption that the passing load is constant, which, however, is not the case, since the number of the operating trains, their positions, etc. are variable. Calculation of the transmission line is accordingly carried out with the help of a small-scale physical model where:
- substations are simulated by constant-voltage sources, complemented by suitable resistors simulating the internal circuits of the stations
- current feeder and return lines are simulated by suitable resistors
- trains are simulated by variable resistors, which can be connected to various points of the line
- suitable measuring instruments give a direct reading (as a function of substation distance and transmission line cross-section) of substation output voltage, total current at each substation, voltage at traction engine switchboards, etc.

This physical model enables the testing and verification of various combinations of transmission conductors, substation distances, etc. and the selection of the optimum solution.

In any case, according to UIC regulations, in the case of 25,000 V, 50 Hz traction, the transmission line voltage, in order to ensure a normal traction engine power supply, should have a maximum value of 27,500 V, a normal value of 25,000 V, a minimum value of 19,000 V, and only a momentary fall to 17,000 V, (200).

13.8.3. Transmission line suspension

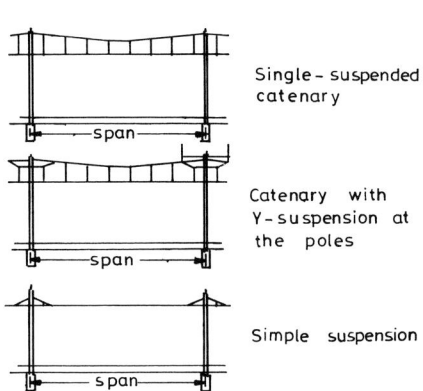

Various transmission line suspension methods are being used (Fig. 13.10), depending mainly on train speed, but also on climatic conditions (wind speed and direction) and on pole spacing. With low speeds (up to 120 km/h), simple suspension is adequate, whereas with medium and high speeds catenary-type suspension is mandatory, (202).

Fig. 13.10. Transmission line suspension methods

13.8.4. Track configuration

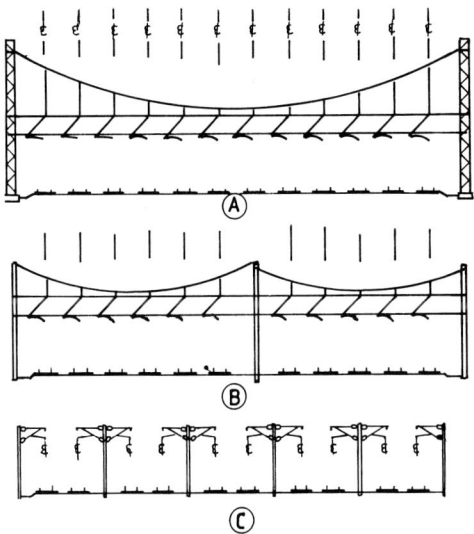

Whenever several tracks are laid parallel (stations, tunnel entrance-exits, bridges, etc.), it is advisable to reconfigure and eliminate certain tracks in order to reduce the total number of tracks to be electrified. Figure 13.11 illustrates such a reconfiguration which allows the elimination of certain tracks and a reduction of the total cost. The reconfiguration should take into consideration future junctions, turnouts and crossings, (197).

Fig. 13.11. Reconfiguration and elimination of tracks before electrification

13.8.5. Power transmission by conductor rail

As mentioned in section 13.5.1, electric power may be supplied to locomotives using either an overhead transmission line or conductor rails (one or two). Conductor rails are mainly used in metros and some suburban railways.

The conductor rail solution (Fig. 13.12) is preferable in the case of very heavy traffic loads, when very large overhead line cross-sections would be necessary. The conductor rail is equivalent to an overhead conductor with a 900 mm^2 cross-section, which in the case of tunnels allows a smaller load gauge and therefore considerable savings, (1).

In the vicinity of level crossings or turnouts, the third rail is interrupted and power supply continuity is ensured by special insulated cables.

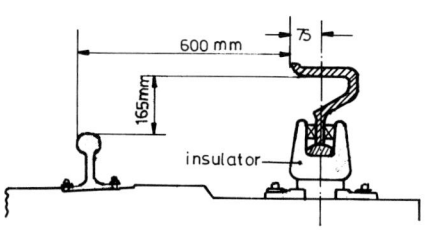

Fig. 13.12. Power supply by conductor rail

Special attention should be paid to safety, possibly covering the conductor rail with insulating plates in personnel working or passage areas. Conductor rails are more sensitive to snow and frost than overhead systems. In some metros (London underground for instance) two conductor rails are used, positive and negative, to avoid earth return on the running rails.

Railway Engineering

13.9. Overhead line supporting poles

13.9.1. Pole material

The poles supporting the overhead line may consist of:
- cast steel
- zinc plated steel
- prestressed concrete
- reinforced concrete.

13.9.2. Pole spacing

Spacing between supporting poles ranges between $50 \div 75$ m depending on the following factors:
- pantograph oscillations
- locomotive transverse motions
- climatic conditions.

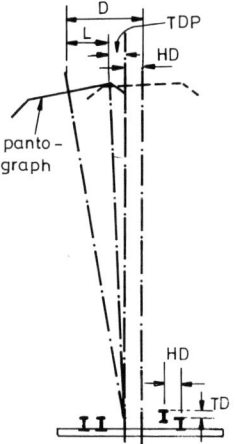

Fig. 13.13. Pantograph oscillations

Fig. 13.13 illustrates the transverse motion D of the pantograph, resulting from the addition of, (197):
- the horizontal defect (see section 11.3.3)
- the transverse defect TD, which is reflected on pantograph motions multiplied by the ratio μ:

$$\mu = \frac{\text{height of overhead line}}{\text{track gauge}}$$

and

$$NTP = TD \cdot \mu$$

- the transverse motion L of the locomotive, depending on the speed of the train, the height of the overhead line, the locomotive suspension springs, etc.

Both longitudinal and transverse pantograph motion have to be calculated in detail. It should be stressed that the primary constraint on maximum train speed (515 km/h in test runs in 1990, see section 1.1.3) is the maximum permissible pantograph oscillations, and to a lesser degree the metal-to-metal (wheel-rail) contact.

13.9.3. Pole foundation

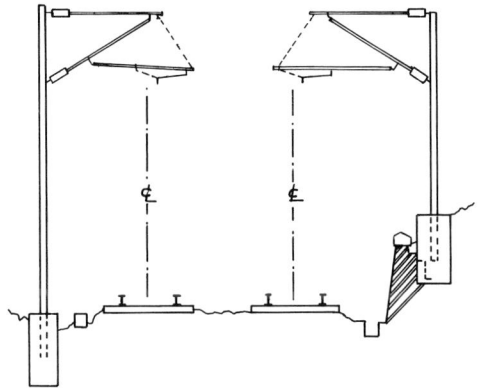

Fig. 13.14. Erection of electric traction poles

When erecting poles for electric traction, special care is required both at the excavation and at the filling-up stages (Fig. 13.14), so as to minimize eventual settlement of the ground.

When poles are erected on solid ground (Fig. 13.15), calculations are based on the moment, (197):

$$M = c \cdot B \cdot L^3$$

with coefficient c depending on soil characteristics.

In the case of poor ground, poles are erected on a small base slab (Fig. 13.16).

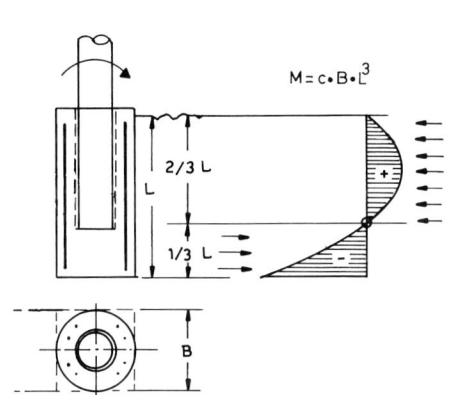

Fig. 13.15. Calculation of pole foundation on good-quality ground

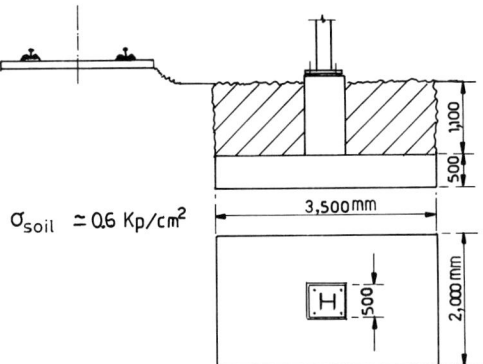

Fig. 13.16. Erection of electric traction poles on poor-quality ground

Railway Engineering

13.10. Substations

13.10.1. Substations feeding direct current systems

Substations supplying DC systems, in addition to stepping the three-phase voltage down, also rectify the AC into DC.

Rectification was initially performed by AC motor - DC generator couples, later superseded by mercury-pool rectifiers and more recently by silicon rectifiers.

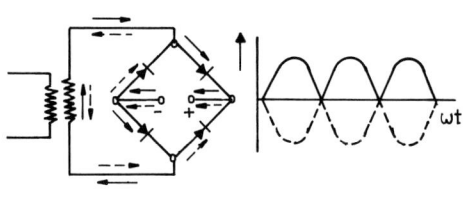

Fig. 13.17. Simplified schematic of a DC substation

A modern substation includes a high-medium voltage transformer with one or two output voltages, and a rectifier assembly (Fig. 13.17). Silicon diodes or thyristors have been used as rectifiers, but since the mid-1980s, they have been replaced by GTO technology (see below section 13.10.3).

13.10.2. Substations feeding alternating current systems

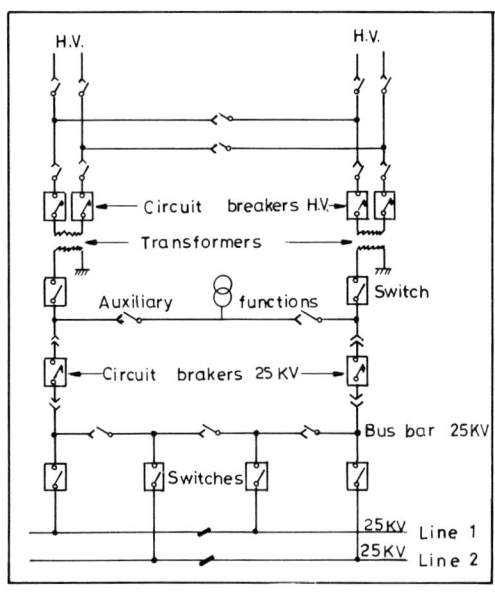

Fig. 13.18. AC substation 25 kV, 50 Hz

In AC substations (Fig. 13.18), only the voltage is being stepped down and therefore substations in this class are simpler than DC substations. AC substation design should take into particular consideration the risk of short-circuiting. This can be prevented by the addition of preventive configurations and devices that limit the risk of short-circuiting.

13.10.3. From thyristor to GTO technology

Thyristors were extensively used until the mid 1980s. The introduction at that time of the Gate Turn Off (GTO) technology (Fig. 13.19) permitted omission of the commutating circuits, thus enabling distinct reduction of load losses (Fig. 13.20).

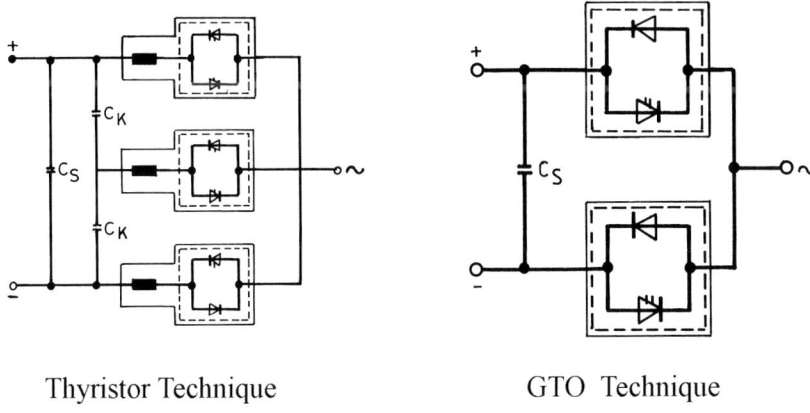

Fig. 13.19. Thyristor and GTO techniques

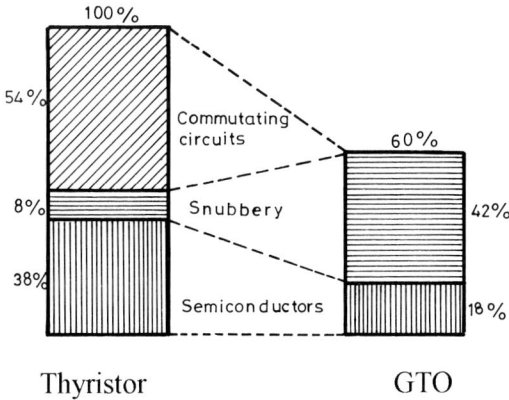

Fig. 13.20. Load losses by thyristor and GTO techniques

Railway Engineering

13.10.4. Central remote control

Today, substations and the systems supplied by them, are remote-controlled and monitored from a central control station, provided with a visual panel showing the lines, substations, and the sections supplied (and therefore controlled) by each substation. Remote control is achieved using a signal code composed of different frequencies. Electronic control circuits in recent years have made possible execution times in the order of 0.3 sec.

13.10.5. Interference of electrical traction on telecommunication systems

In addition to power transmission lines (in the case of electric traction), railway telecommunication cables are also running (usually underground) alongside railway tracks. In order to prevent interference between power and telecommunication cables, voltages induced in the telecommunications network should precisely be calculated.

Problems may also arise where traction power cables intersect with lines of the public power network.

13.11. Synchronous and asynchronous motors

Electric motors may be classified into the following three general categories (Fig. 13.21):

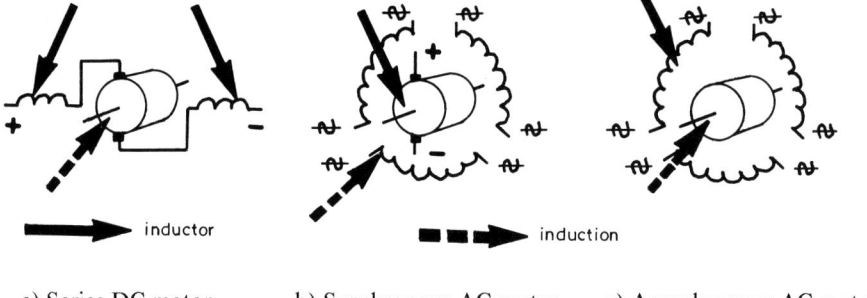

a) Series DC motor   b) Synchronous AC motor   c) Asynchronous AC motor

Fig. 13.21. The three classes of electric motors

- *Direct current motors.* The inductor is fixed (stator) and carries DC. Induction takes place between the stator and the moving part or rotor, which is supplied with DC through brushes, so that the rotor windings carry alternating current. Motor speed is adjusted by varying the DC voltage applied to the motor as well as by varying the induced magnetic field. The direction of rotation is reversed by inverting the inductor connections (polarity reversal).

- *Synchronous motors.* The inductor is rotating (rotor) and carries DC. Induction takes place between the rotor and the fixed part (the stator) which carries three-phase AC. Rotation speed is adjusted by varying the frequency of the three-phase alternating current. The sense of rotation is reversed by reversing the AC phase sequence.

- *Asynchronous motors.* The inductor is fixed (stator) and carries three-phase AC. Induction takes place between the stator and the rotating part (rotor) which carries three-phase AC. Speed is adjusted by varying the three-phase AC frequency. The sense of rotation is reversed by reversing the inductor phase sequence.
  Asynchronous motors offer the following advantages:
  - Lighter weight, about half compared to synchronous motors of the same power.
  - Higher efficiency and torque.
  - Less track loading.

Most electric locomotives in current service use direct-current motors. The French Railways, for instance, employ electric locomotives of the BB series, manufactured by Alsthom, weighing 90 tonnes, with a power of 4,400 kW and speed 160 km/h; the Swedish Railways use R/C-series locomotives manufactured by ABB; the German Railways use E 181.2-series locomotives manufactured by Krupp with a power of 3,300 kW and a speed of 160 km/h. Examples of asynchronous motors are the German ICE, the high-speed Eurostar Paris-London rail link, etc.

Synchronous and asynchronous motors are practically equivalent concerning power. They are more efficient than direct-current motors because of their greater speed of rotation. Asynchronous technology is expanding rapidly, in spite of complicated electronic systems of command. However, a locomotive with a synchronous motor run in 1990 the speed of 515 km/h (see section 1.1.3). Selection between synchronous and asynchronous motors must be based on analysis of the purchase, operation and maintenance cost of each.

13.12. Electric locomotive maintenance - depot

A critical factor for the good operation of the rolling stock is efficient and in time maintenance. Maintenance must be preventive and be based on the following principles:
- specialization of staff and equipment
- in-time scheduling of maintenance sessions

Railway Engineering

- appropriate mechanical and computer equipment for the accurate monitoring of any deficiencies
- continuous control and evaluation of results
- reduction of cost

For electric traction engines, various routine inspections and maintenance must be performed: two-day inspection, weekly inspection, monthly technical inspection, two-month maintenance, four-month maintenance, yearly maintenance, general overhaul every 10 years, general overhaul every 20 years.

Maintenance up to the four-month level can be performed in the local depot. Beyond this level, repairs are conducted at a maintenance facility.

List of References

GENERAL REFERENCES

1. Alias, J. (1984), *La Voie Ferrée - Tome 1: Techniques de Construction et d'Entretien*, Eyrolles, Paris.
2. Alias, J. (1993), *La Voie Ferrée - Tome 2: Signalisation*, Ecole Nationale des Ponts et Chaussées (ENPC), Paris.
3. Alias, J. (1993), *La Voie Ferrée - Tome 3: Exploitation Technique et Commerciale*, ENPC, Paris.
4. Esveld, C. (1989), *Modern Railway Track*, The Netherlands.
5. *World Congress on Railway Research* (1994), International Congress, Paris.
6. Oliveros Rives, F., Rodriguez Mendez, M., Megia Puente, M. (1983), *Tratado de Explotación de Ferrocarriles*, Editorial Rueda, Madrid.
7. Railway Industry Association of Great Britain (1983), *Track Course*, London.
8. Fiedler, J. (1991), *Grundlagen der Bahntechnik*, Werner-Verlag, Düsseldorf.
9. The Permanent Way Institution (1993), *British Railway Track*, 6th Edition, Echo Press Ltd, Loughborough.
10. H.M. Stationery Office (1992), *Railway Construction and Operation Requirements - Structural and Electrical Clearances*, London.
10a. Murthy, T.K., Mellitt, B., Brebbia, C.A. (1994), *Computers in Railways*, Computational Mechanics Publications, Southampton.

REFERENCES PER CHAPTER

CHAPTER 1

11. World Bank (1982), *The Railway Problem*, Washington.

12. Union Internationale des Chemins de Fer (UIC) (1980), *Le Chemin de Fer d'Aujourd'hui et de Demain*, Paris.
13. UIC (1994), *The Role of the Railways in the Development of the Economy*, International Seminar, N. Delhi.
14. European Conference of Ministers of Transport (ECMT) (1986), *European Dimension and Prospects of the Railways*, International Seminar, Paris.
15. ECMT (1990), *Prospects for East-West European Transport*, International Seminar, Paris.
16. ECMT, Round Table 47 (1980), *Scope for Railway Transport in Urban Areas*, Paris.
17. ECMT (1985), *Improvements in International Railway Transport Services*, Paris.
18. ECMT (1986), *High Speed Traffic on the Railway Network of Europe*, Paris.
19. ECMT (1989), *Rail Network Cooperation in the Age of Information Technology and High Speed*, Paris.
20. UIC, *Railway Statistics, 1985-1994*, Paris.
20a. ECMT (1993), *Transport Evolution 1970-1992*, Paris.
21. Commission of the European Communities (1990), *The European High-speed Train Network*, Brussels.
22. Metzler, J.M. (1980), *Les Grandes Vitesses Ferroviaires*, ENPC.
23. Profillidis, V. (1985), 'High-Speed Trains', *Technika Chronika (Scientific Journal of Greek Engineers)*, Vol.5, No 3, Athens.
24. *Modernization of Railway and Airway Transport - The Impact of Liberalization* (1994), International Conference, Democritus Thrace University, Xanthi.
24a. ECMT (1995), *Why Do We Need the Railways*, International Seminar, Paris.
25. Gohlke, R. (1988), 'The Future of the European Railways', *Rail International*, August - September 1988.
26. Bouley, J. (1983), 'Changing Course in Changing Times', *International Railway Journal*, February 1983.
27. Amatore, P. (1990), 'The Future of the Railways in an Integrated European System', *Rail International*, April 1990.
28. Batisse, F. (1988), 'La Mobilité Ferroviaire dans le Monde', *Révue Générale des Chemins de Fer* (RGCF), September 1988.
29. Profillidis, V. (1990), 'Present Status and Future Prospects of Greek Railways - An Analysis of a Railway Network in a Difficult Situation', *Transportation Planning and Technology*, Vol. 14.
30. RGCF (1992), *Le TGV Nord*, RGCF, January-February 1992.
31. Alston, Liviu (1984), *Railways and Energy*, World Bank Editions.
32. UIC (1993), *The Railways - An Indispensable Part of the European Transport System*, Paris.

List of References

33. ENPC (1989), *L'Europe des Transports et des Communications*, Séminaire Internationale, Paris.
34. Estival, J.-P., Profillidis, V. (1985), 'For a New Strategy of the European Rail Networks', *Rail International*, July 1985.
35. Profillidis, V. (1986), 'Railway Infrastructure of the Networks of the Peripheral Countries of the European Economic Community', *Rail International*, December 1986.
35a. RGCF (1992), *Le Tunnel sous la Manche*, December 1992.
36. Brubel, H. (1977), 'Die Technische Gestaltung des Newbaustrecken der Deutschen Bundesbahn', *Die Eisenbahningenieur*, Vol. 1, No 1.
37. Misiti, L. (1978), 'La Direttissima Roma-Firenze', *Ingegneria Ferroviaria*, N.1.
38. Devaux, P. (1989), *Les Chemins de Fer*, Presses Universitaires de France.
39. Hochbruck, H. (1984), 'Perspektiven des Schudlverkehrs, Magnet-und/ oder Rad/Schiene-Technik', *Eisenbahntechnische Rundschau*, No 7/8, Darmstadt.
40. Roumeguère, Ph. (1985), 'Les Installations Fixes du TGV Deux Ans après leur Mise en Service', *Rail International*, Vol. 8/9.
41. Profillidis, V. (1994), *Transport Economics*, Democritus University Press.
42. UIC (1991), *Pour une Transformation du Système Ferroviaire International dans le Cadre d'une Politique Nouvelle des Transports en Europe*, Paris.
43. ECMT (1994), *Internalizing the Social Costs of Transport*, Paris.
43a. Estival, J.-P. (1988), *Rapport de Mission sur les Méthodes des Coûts Marginaux utilisés au sein des Réseaux du Groupe des Dix*, CEE-UIC.
44. RGCF (1991), *Dix Ans de TGV*, October 1991.
45. Roumeguère, Ph. (1994), Le Schéma Directeur Français des Liaisons à Grande Vitesse', *RGCF*, June-July 1994.
46. Brand, M.M., Lucas, M.M. (1989), 'Operating and Maintenance Costs of the TGV High Speed Rail System', *American Society of Civil Engineers (ASCE), Journ. of Transp. Eng.*, Vol. 115, No 1.
46a. Lammich, K. (1994), 'Deutschland nach dem Tarifaufhebungsgesetz: Was bleibt übrig von der Kontrollierten Verkehrsmarketordnung', *Deutscher Verkehrs*, No 1-2, Hamburg.
46b. Pintag, G. (1989), 'Capital Cost and Operations of High Speed Rail System in West Germany', *ASCE, Journ. of Transp. Eng.*, Vol. 115, No 1.
46c. Profillidis, V. (1987), 'A Methodology of Quantification of the Public Benefit that the Railways offer to the Society and a New Approach for the Appreciation of the Management of the Railway Undertaking', *XVII Panamerican Railway Congress Association*.
47. Profillidis, V. (1990), 'Light Rail Technologies in the 1990s', *International Conference on Electric Transport*, November 1990, Basel.
47a. Profillidis, V. (1995), 'Light Rail Transit Systems: Present Trends and Future Prospects', *Journ. of Light Rail Transit Association*, Jan. 1995, London.

47b. ECMT (1994), *Light Rail Transit Systems*, Paris.
47c. ECMT (1992), *Guided Transport in 2040*, Paris.
48. Profillidis, V. (1991), 'Combined Transport between Greece, Europe and the Middle East-Present Trends and Future Prospects', *International Conference, University of Trieste*, September 1991.
49. ECMT (1986), *The Cost of Combined Transport*, Paris.
49a. Institute of Transport, Univ. of Trieste (1994), *Logistics and Transport in Europe for the Year 2000*, International Seminar, Trieste.
49b. ECMT (1993), *Possibilities and Limitations of Combined Transport*, Round Table 91, Paris.
50. Legrand, P., Chubaneix, J.-P. (1994), 'Sirène, le Serveur International pour la Réalisation des Echanges Normalisés en Europe', *RGCF*, April 1994.
51. Rigaud, G., Ousten, J. (1992), 'Chemins de Fer et Energie', *RGCF*, December 1992.
52. Auzannet, P., Bellaloum, Ad. (1993), 'Le Coût des Transports pour la Collectivité', *RGCF*, March 1993.
53. Raschbichler Hg. (1992), 'Die Magnetschnellbau Transrapid - Ein Neues Verkehrsystem für des Personen - und Gütertransport', *Zeitschrift für Eisenbahnwesen und Verkehrestechnik*, No 8-9, Berlin.

CHAPTER 2

54. Sauvage, R., Richez G. (1978), 'Les Couches d' Assise de la Voie Ferrée', *RGCF*, December 1978.
55. Profillidis, V. (1983), *La Voie Ferrée et sa Fondation-Modélisation Mathématique*, Ph.D. Thesis, Ecole Nationale des Ponts et Chaussées, Paris.
56. Prud' homme, A. (1970), 'La Voie', *RGCF*, Paris, Janvier 1970.
56a. Bonnett, C.F. (1992), 'Trackwork for Lightweight Railways', *Proceedings of the Institution of Civil Engineers*, Part 1.
57. Bendat, J., Piersol, A. (1971), *Random Data: Analysis and Measurement Procedures*, Wiley, New York.
58. Ohyama, T. (1992), 'Adhesion Characteristics of Wheel-Rail System and its Control at High Speeds", Quarterly Report, *RTRI*, Tokyo, Vol. 33, No 1.
59. Organisme des Recherches et d' Essais (ORE), Committee C152, Report (RP) No 2 (1983), *Preliminary Study concerning the Application of the Mathematical Methods for Characterizing the Vehicle - Track Interaction*, Utrecht.
60. Deutsche Bundesbahn (1993), *Bundesbahn-Zentralamt München*, DB - TL 918235.
61. ORE, C116, RP 10 (1981), *Study of Optimum Rail Inclination and Gauge Related to Wheel Profiles adapted to Wear*, Utrecht.

List of References

61a. Kalker, J. (1967), *On the Rolling Contact of Two Elastic Bodies in the Presence of Dry Friction*, Ph.D. Dissertation, Delft.
62. Esveld, C. (1978), *Spectral Analysis of Track Geometry for Assessing the Performance of Maintenance Machines*, ORE, DT 77, Utrecht.
63. ORE, D 161, RP 4 (1987), *The Dynamic Effects due to Increasing Axle Loads from 20 to 22.5 t*, Utrecht.
63a. Kaess, G., Theiss, H. (1984), 'Der Oberbau auf den Neubaustrecken der DB', *Eisenbahningenieur*, Vol. 9.

CHAPTER 3

64. Profillidis, V., Kouparoussos A. (1984), 'Mechanical Behaviour of the Railway Subgrade', *KEDE, Scient. Bulletin of the Ministry of Public Works of Greece*, Vol. 3-4, Athens.
65. UIC, Fiche 719 (1980), *Ouvrages en Terre et Couches d'Assise Ferroviaires*, Paris.
66. UIC, Question 7H14 (1978), *Adaptation de la Plate-forme dans l'Optique des Circulations à Grande Vitesse et de l'Augmentation de la Charge par Essieu*, Paris.
67. Deutsche Bundesbahn (1974), *Vorläufige Richtlinien für Plannung und Herstellung der Erdbauwerke von Strecken mit hohen Geschwingdigkeit*, OS 836/2.
68. Hartmark, H. (1979), 'Frost Protection of Railway Lines', *Engin. Geology*, Vol. 13, Amsterdam.
69. Ayres, D. (1961), 'The Treatment of Unstable Slopes and Railway Track Formations', *Journ. of the Soc. of Engineers*, Vol. 52, No 4, London.
70. Sugiyama, T., Okada, K., Muraishi, H. (1993), 'Development of Method for Predicting Railway Embankment Collapse due to Rainfall Based on Damage Examples', Quarterly Report, *RTRI*, Tokyo, Vol. 34, No 4.
71. Taillé, J.-Y. (1992), 'TGV Nord et Géologie', *RGCF*, Jan.-Febr. 1992.
72. Tirant, P., Sarda, J. (1965), 'Chargements Répétés des Sols Fins Compactés et Non Saturés', *Bull. de Liais. des Labor. des Ponts et Chaussées*, (LCPC), July-August 1965.
73. Sauvage, R., Langlade, J. (1981), 'L' Utilisation des Géotextiles dans les Plates-formes Ferroviaires de la SNCF', *RGCF*, July-August 1981.
74. Profillidis, V. (1985), *Geotextiles - Mechanical and Hydraulic Behaviour - Applications*, Textbook, Thessaloniki.
75. Rankilor, D. (1981), *Membranes in Ground Engineering*, John Wiley.
76. Perrier, H. (1983), *Sol Bicouche Renforcé par Géotextile*, LCPC, Paris.
77. ORE, D117, RP15,16 (1981), *Filtration et Drainage*, Utrecht.
78. SNCF (1982), *Ouvrages en Terre Armée*, Paris.
79. République Française, Ministère des Transports (1979), *Ouvrages en Terre Armée-Recommandations et Règles d' Art*, Paris.

79a. Caillou, J., Vallet, D., Cervi, G. (1994), 'La Voie sans Ballast - Vers une Solution pour les Grandes Lignes', *RGCF*, June-July 1994.
80. Rowe, K. (1984), 'Reinforced Embankments: Analysis and Design', ASCE, *Journal of the Geotechnical Engineering Division* (JGED), Vol. 110, No 2.

CHAPTER 4

81. Eisenmann, J. (1977), *Die Schiene als Träger und Fahrbahn*, Verlag Ernst, Berlin.
81a. Zimmermann, (1941), *Die Berechnung des Eisenbahnoberbaues*, 3d Edition, Wilhelm Ernst und Sohn, Berlin.
82. Hill, R. (1950), *The Mathematical Theory of Plasticity*, Oxf. Univ. Press.
83. Chang, C., Adegoke, C., Sellig, F. (1980), 'Geotrack Model for Railroad Track Performance', *ASCE, J.G.E.D.*, Vol. 106, No GT 11.
84. Desai, C., Siriwardane, H. (1982), 'Numerical Models for Track Support Structures', *ASCE, J.G.E.D.*, Vol. 100, No GT3.
85. Zienkiewicz, O. (1980), *The Finite Element Method in Engineering Science*, McGraw-Hill.
86. Imbert, J. (1979), *Analyse des Structures par Eléments Finis*, Editions Cepadues, Toulouse.
87. Zienkiewicz, O., Valliapan, S., King, I. (1969), 'Elastoplastic Solutions of Engineering Problems. Initial Stress-Finite Element Approach', *Int. Journ. of Num. Meth. in Engin.*, Vol. 1.
88. Salençon, J., Halphen, B. (1984), 'Elasto-plasticité', ENPC.
89. Ecole Polytechnique (1972), *Plasticité et Visco-plasticité*, Seminar, Paris.
90. Drucker, D. (1951), 'A More Fundamental Approach to Plastic Stress-Strain Relations', *Proceedings, 1st U.S. Nat. Congr. Appl. Mech.*
91. Profillidis, V. (1986), 'Applications of Finite Element Analysis in the Rational Design of Track Bed Structures', *Computers and Structures*, Vol. 22, No 3.
92. Profillidis, V., Humbert, P. (1986), 'Etude en Elastoplasticité par la Méthode des Eléments Finis du Comportement de la Voie Ferrée et de sa Fondation', *Bull. de Liaison des Laboratoires des Ponts et Chaussées*, Vol. 141.
93. Profillidis, V. (1985), 'Three-Dimensional Elasto-Plastic Finite Element Analysis for Track Bed Structures', *Civil Engineering for Practicing and Design Engineers*, Vol. 4, No 9.
94. Lopez Pita, A., Oteo Mazo, C. (1978), 'Análisis de la Deformabilidád de Una Via Férrea Mediante el Método de Elementos Finitos', *AIT*, No. 15.
95. Profillidis, V., Poniridis, P. (1990), 'Non-linear Analysis of Metropolitan Railway on Reinforced Concrete Slab', *Scientific Bulletin of the Ministry of Public Works of Greece*, Vol. 105-106, Athens.

96. Profillidis, V., Poniridis, P. (1986), 'The Mechanical Behaviour of the Sleeper-Ballast Interface', *Computers and Structures*, Vol. 24, No 3.
97. Prud'homm, A. (1976), 'Les Problèmes que Pose pour la Voie la Circulation des Rames à Grande Vitesse', *RGCF*.
98. ORE, D71, RP 9, 10 (1978), *Stress in the Track, Ballast and the Subgrade under the Action of Repeated Loading*, Utrecht.
99. ORE, D 117, RP 18,25,27,28,29 (1984), *Optimum Adaptation of the Conventional Track to Future Traffic*, Utrecht.
100. Deutsche Bundesbahn (1978), *Hertstellung von Planumsschutzschichten aus Korngemischen*, München.
101. Frederick, C. (1987), 'Vibrations in Ground: Railway Induced Ground Vibrations', *Rail International*, October 1987.
102. Girardi, L. (1981), 'Propagation des Vibrations dans les Sols Homogènes ou Stratifiés', *Inst. Techn. du Bat. et des Trav. Publ.*, No 397.
103. Gutowski, T., Dym, C. (1976), 'Propagation of Ground Vibration: a Review', *Journ. of Sound and Vibration*, No 49.
104. Clement, H. (1994), 'Les Voies Ferrées de Metro et la Protection de l'Environnement en Milieu très Urbanisé', *RGCF*, April 1994.
104a. Wayson, R.L., Bowlby, W. (1989), 'Noise and Air Pollution of High-Speed Rail Systems', *ASCE, Journ. of Transp. Eng.*, Vol. 115, No 1.
104b. Japan Environment Agency (1985), 'Further Measures for Achieving Environmental Quality Standards for Shinkansen Railway Noise', *Jap. Envir. Sum.*, Vol. 13.
105. ORE, C137, RP 12 (1981), 'Railway Noise: Measurements of the Running Noise caused by Trains on Different Types of Bridges', Utrecht.
106. The Institution of Civil Engineers (1984), *Track Technology*, Conference Proceedings, Nottingham.
106a. Shenton, M. (1987), *Track Standards for High Speed Trains*, ORE Colloquium, Arrezo (Italy).
107. Alias, J. (1982), 'Sans Bonnes Voies, pas de Chemin de Fer Sûr', *Le Rail et le Monde*, No 17.
107a. Zicha, J.H. (1989), 'High-Speed Rail Track Design', *ASCE, Journ. of Transp. Eng.*, Vol. 115, No 1.
108. Wiley, R. (1975), *Advanced Engineering Mathematics*, McGraw-Hill.

CHAPTER 5

109. Profillidis, V. (1991), 'Mechanical Behaviour of the Rail', *Professor G. Nitsiotas's Honorary Volume*, University of Thessaloniki.
110. Dang Van, K., Gence, P. (1978), 'Evolution des Critères de Fatigue-Application au cas des Rails, *RGCF*, December 1978.
111. Eisenmann, J. (1970), 'Stress Distribution in the Permanent Way due to Heavy Axle Loads and High Speeds', *AREA*, Vol. 71.

112. Fowler, G. (1976), *Fatigue Crack Initiation and Propagation in Pearlitic Rail Steels*, Ph.D. Thesis, Univ. of California.
113. ORE, D 71, RP 2 (1966), *Stress Distribution in the Rails*, Utrecht.
114. ORE, D 117, RP 3 (1973), *Rail Behaviour in Relation to Operation Conditions*, Utrecht.
115. Lévy, D. (1989), 'Conception d'un Système de Mesure et d'Analyse de l'Usure Ondulatoire des Rails', *RGCF*, May 1989.
116. Mair, R., Groenhout, P. (1981), 'Croissance des Défectuosités Transversales dues à la Fatigue dans le Champignon des Rails de Chemin de Fer', *Rail International*, February 1981.
117. Sauvage, R., Amans, F. (1969), 'Railway Track Stability in Relation to Transverse Stresses exerted by Rolling Stock - A Theoretical Study of Track Behaviour', *Rail International*, Nov. 1969.
118. Sauvage, G., Pascal, P. (1990), 'Nouvelle Méthode de Calcul des Efforts Dynamiques entre les Roues et les Rails', *RGCF*, December 1990.
119. Sperring, D., Squires, J. (1983), 'Rail Wear and Associated Problems', *British Railways Track Course*.
120. Profillidis, V. (1986), 'Continuous-Welded Rail', *Bulletin of Greek Civil Engineers*, No.172, Athens.
121. Timoshenko, S., Langer, B. (1932), 'Stress in Railroad Track', *ASME*, Vol. 54.
121a. Profillidis, V. (1995), 'A Unitaleral Theory Approach of the Rail-Sleeper and Sleeper-Ballast Contact', to appear in *Computers and Structures*.
122. Panagiotopoulos, P.D. (1985), *Inequality Problems in Mechanics and Applications - Convex and Non-Convex Energy Functions*, Birkhäuser Verlag.
122a. Panagiotopoulos, P.D. (1993), *Hemivariational Inequalities - Applications in Mechanics and Engineering*, Springer.
123. Tassily, E. (1987), 'Propagation des Ondes de Flexion dans la Voie Ferrée considerée comme un Milieu Périodique', *RGCF*, March 1987.
124. Tounend, P. (1980), 'Analyse de la Probabilité et Coût des Défauts en Forme de Tache Ovale dus à la Fatigue des Voies en Alignement et en Courbe dans des Conditions de Fortes Charges par Essieu', *Rail International*, July-August 1980.
124a. Alfelor, R.M., Mc Neil, S. (1994), 'Heuristic Algorithms for Aggregating Rail-Surface-Defect Data', *ASCE, Journ. of Transp. Eng.*, Vol. 120, No 2.
125. Yasojima, Y., Machii, K. (1965), 'Residual Stresses in the Rail', *Permanent Way*, No 26, Vol. 8, Society of Japan.
126. UIC (1979), *Catalogue of Rail Defects*, Paris.
127. Edel, K., Ortmann, R. (1990), 'Fracture-Mechanical Characteristics of Rail Materials', *Rail International*, August-September 1990.
128. ORE, DT 119 (1981), *Etude de l'Influence du Chargement des Essieux sur le Tassement de la Voie*, Utrecht.

List of References

129. ORE, D 141, RP 2 (1979), *Influence at the Track of an increase of Mass Axle from 20 to 22.5 t*, Utrecht.
130. ORE, D 141, RP 1 (1979), *Statistical Study of the Evolution of Rail Defects in Relation to the Medium Axle Mass*, Utrecht.
131. Zarembski, A.M. (1979), 'Effect of Rail Section and Traffic on Rail Fatigue Life', *American Railway Engineering Association, Bulletin* 673, Vol. 80.
132. Matsuura, A. (1992), 'Dynamic Interaction of Vehicle and Track', *Quarterly Report*, RTRI, Vol. 33, No 1, Tokyo.
133. Orringer, O., Morris, J.M., Steele, R.K. (1984), 'Applied Research on Rail Fatigue and Fracture in the United States', *Theoretical and Applied Fracture Mechanics*, Vol. 1.

CHAPTER 6

134. FIP (Fédération Internationale de la Précontrainte) (1987), *Concrete Railway Sleepers*, Thomas Telford Editions, London.
135. ORE, D71, RP8 (1973), *Load Distribution under the Sleeper*, Utrecht.
136. Buekette, J. (1983), 'Concrete Sleepers', *Track Course*, RIA, London.
136a. American Railway Engineering Association (1982), *Concrete Ties*.
137. SATEBA (1992), *Twin-Block Railway Sleepers*, Paris.
137a. Hodgson, W.H. (1983), 'Steel Sleepers', *Track Course*, RIA, London.
138. ORE, D71 (1973), *Sollicitations de la Voie, du Ballast et de la Plateforme*, Utrecht.
139. Bonewitz, W., Führer, G. (1992), 'Eisatz von Elastomeren bei Shienenbefestigung bei Eisebahnen und Nahverkehrsbahnen', *Die Bundesbahn*, No 3, Darmstadt.
140. Lindsey, D. (1983), 'Rail Track Fastenings', *Track Course*, RIA, London.
141. ORE, D11, RP1 (1974), *Methods of Fastening Rails to Sleepers*, Utrecht.
141a. Watanabe, J. (1980), 'Engineering of Rail Fastening', *Japan. Rail. Eng.*, Vol. 19, No 45.
142. Profillidis, V., 'The Mechanical Behaviour of the Railroad Tie', *ASCE, Journ. of the Stuctural Engin. Div.*, (under print), 1995.
143. Sauvage, R., Errieau, J. (1970), 'Les Poses de Voie sans Ballast', *RGCF*, March 1970.
144. Brown, J. (1983), 'Continuous Slab Track', *Track Course*, RIA, London.
144a. Squires, J.H., Sperring, D.G. (1983), 'Theory and Development of Resilient Pads', *Track Course*, RIA, London.
144b. Eisenmann, J. (1977), *Investigations into Behaviour of Plastic Pads*.

CHAPTER 7

145. Profillidis, V. (1988), 'Mechanical Behaviour of the Railroad Ballast', *1st Panhellenic Congress of Geotechnical Mechanics*, Athens.

146. Raymond, G., Davies, J. (1978), 'Triaxial Tests on Dolomite Railroad Ballast', *ASCE, Journ. of the Geotechn. Engin. Div.*, Vol. 104, No GT6.
147. Brown, S. (1978), 'Repeated Load Testing of a Granular Material', *ASCE, Journ. of the Geotechn. Engin. Div.*, Vol. 104, No GT6.
148. Lopez Pita, A. (1977), 'Analyse de la Déformabilité du Ballast au moyen d'Essais en Laboratoire', *Associación de Investigación del Transporte*, Madrid.
149. Hartmark, H. (1979), 'Frost Protection of Railway Lines', *Engineering Geology*, Vol. 13.
150. Gray, P.S. (1983), 'Structural Requirements and Specifications of Ballast', *Track Course*, RIA, London.
151. ORE, D117, PR5 (1974), *Deformation of Track Ballast under Repeated Loading*, Utrecht.
152. SNCF (1979), *Constitution de la Voie Courante*, Paris.
153. ORE, D117, RP28 (1983), *Tables for the Behaviour of the Track-Subgrade System*, Utrecht.
153a. Wilmott, D.J. (1983), 'New Track Construction', *Track Course*, RIA, London.

CHAPTER 8

154. Profillidis, V. (1987), 'Parametric Analysis of Transverse Track Resistance and Application to the Design of the Ballast Section', *KEDE Scient. Bulletin of the Ministry of Public Works of Greece*, Vol.1-2, Athens.
155. ORE, C138, RP8 (1984), *Permissible Maximum Values for the Y- and Q-Forces and Derailment Criteria*, Utrecht.
156. Moreau, A. (1987), 'La Verification de la Securité contre le Déraillement', *RGCF*, April 1987.
157. Amans, F., Sauvage, R. (1969), 'La Stabilité de la Voie vis-à-vis des Efforts Transversaux Exercés par les Véhicules', *Annales des Ponts et Chaussees*, Vol. 1.
157a. Erchkov, O.P., Kartzev, V.J. (1980), 'Recherches Théoriques et Expérimentales sur les Mouvements des Véhicules Ferroviaires Circulant à une Vitesse de 200 km/h et Exigences Relatives à l'Entretien des Lignes à Grande Vitesse', *Rail International*.
158. ORE, C138, RP5 (1980), *Effect of Train Speed on the Permissible Maximum Value of Load $\Sigma Y=S$ from the Point of View of Track Displacement*, Utrecht.
159. ORE, D117, RP8 (1976), *Influence of Various Measures at the Lateral Resistance of an Unloaded Track*, Utrecht.
160. ORE, C138, RP7 (1982), *Influence des Variations Oscillatoires de la Charge d'Essieu sur la Valeur Maximale Admissible de l'Effort Transversale du Point de Vue du Déripage de la Voie*, Utrecht.

161. ORE, B55, RP8 (1983), *Prevention of Derailment of Goods Wagons on Distorted Tracks*, Utrecht.

CHAPTER 9

162. UIC, 703R (1989), *Layout Characteristics for Lines Used by Fast Passenger Trains*, Paris.
163. Bourguet, A., Joly, R. (1993), 'Rail Vehicle Operation on Curves with a Constant Radius and with a Variable Radius', *Rail International*, December 1993.
163a. Taille, J.-Yv. (1990), 'Naissance d'une Ligne Nouvelle-Les Etudes de Tracé', *RGCF*, Paris.
164. Gubar, J. (1990), 'Railway Transition Curve Planning Methods', *Rail International*, April 1990.
165. Esveld, C. (1991), 'Digital Assessment of Geometrical Track Quality', *Rail International*, April 1991.
166. Hofer, M. (1964), *Absteken von Kreisbogen*, Springer.
166a. Busby, R.H., Drake, D.G.H. (1983), 'Feasibility Studies and Outline Design', *Track Course*, RIA, London.
166b. Adler, H. (1987), *Economic Appraisal of Transport Projects*, The World Bank, Washington.
166c. Roe, M. (1987), *Evaluation Methodologies for Transport Investment*, Avebury, Aldershot.

CHAPTER 10

167. Oeconomos, J. (1987), 'Les Nouveaux Appareils de Voie UIC 60 de la SNCF', *RGCF*, March 1987.
168. Bourda, A. (1991), 'Un Système d'Information pour les Postes d'Aiguillage et de Circulation', *RGCF*, January 1991.
169. Lugg, P. (1983), 'Crossings and Turnouts', *Track Course*, RIA, London.
170. ORE, C138, RP8 (1984), *Permissible Maximum Values for the Y- and Q-Forces and Derailment Criteria*, Utrecht.
171. Deutsche Bundesbahn (1988), Merkblatt für den Entwurf von Gleisanschlüssen, Frankfurt.

CHAPTER 11

171a. Profillidis, V. (1986), 'Basic Principles for the Track Maintenance Works', *Technika Chronika*, Vol. 6, Issue 3, Athens.
172. Janin, G. (1982), 'La Maintenance de la Géometrie de la Voie', *RGCF*, June 1982.

173. ORE, D117, RP2, RP7 (1973), *Etude de l'Evolution du Nivellement en Fonction du Trafic et des Paramètres d'Armement*, Utrecht.
174. ORE, C9, RP9, 'Tolérances en Service Admises dans la Superstructure de la Voie en Relation avec son Etat et la Marche des Véhicules', Utrecht.
175. Lewis R. (1983), 'Track Recording Machines', *Track Course*, RIA, London.
176. Collins R. (1983), 'Heavy Duty High Speed Lines', *Track Course*, RIA, London.
177. Thomas, Cl., Vallée, Cl. (1992), 'TGV Nord - L'Entretien des Installations Fixes', *RGCF*, January-February 1992.
178. Tardieu Gaspar J.E. (1983), 'Algunas Consideraciónes sobre la Renovación de Via', *AIT*, No 53, Madrid.
179. Sasama, H. (1994), 'Maintenance of Railway Facilities by Continuously Scanned Image Inspection', *Japan. Rail. Eng.*, Vol. 33, No 2, Tokyo.
180. Renaux, Ph., Bergeon, D. (1994), 'La Voiture de Mesure de la Géometrie de la Voie de la RATP', *RGCF*, Mai 1994.
180a. Nozawa, D. (1980), 'Present Condition of Track Maintenance', *Japan. Rail. Eng.*, Vol. 20, No 3, Tokyo.
180b. Waghorn, D.W. (1983), 'Weed Control', *Track Course*, RIA, London.

CHAPTER 12

181. Metzler, J.-M. (1989), *Généralités sur la Traction*, ENPC, Paris.
182. ABB (1992), *Traction Vehicle Technic for All Applications*, Mannheim.
183. SNCF (1988), '*La Dynamique du Mouvement des Trains*', Paris.
184. Bianchi, C. (1980), 'Fenomeni Aerodinamici della Marchia Veloce in Galleria', *Tecnica Professionale*, February 1980, Roma.
184a. Moritoh, Y., Zenda, Y. (1994), 'Aerodynamic Noise of High speed Railway Cars', *Japan. Rail. Eng.*, Vol. 34, No 1, Tokyo.
185. Guicheu, C. (1982), 'La Résistance à l'Avancement', *RGCF*, January 1982.
186. Lacôte, F. (1992), 'The Limits of the Wheel-Rail Contact System', *Rail International*, June-July 1992.
186a. UIC, *Codes for Braking*: 540 V, 544-1, 543 VE, Paris.
186b. Wende, D. (1983), *Fahrdynamik*, Transpress VEB Verlag für Verkehrswesen.
187. Boiteux, M. (1990), 'Influence de la Vitesse et de Différents Paramètres Constructifs sur l'Adhérence en Freinage', *RGCF*, July-August 1990.
188. Leluan, A. (1990), 'Méthodes d'Essais de Fatigue et Modèles d'Endommagement pour les Structures de Véhicules Ferroviaires', *RGCF*, December 1992.
189. Edel, K.-O., Shaper, M. (1992), 'Fracture Mechanics Fatigue Resistance Analysis of the Crack - Damaged Tread of Overbraked Solid Railway Wheels', *Rail International*, November 1992.

190. Joly, R. (1988), 'Circulation d'un Véhicule Ferroviaire en Alignement et en Courbe - Bogie à Essieux Auto-Orientés', *Rail International*, April 1988.
191. Pyrgidis, Ch. (1990), *Etude de la Stabilité Transversale d'un Véhicule Ferroviaire en Alignement et en Courbe-Nouvelles Technologies des Bogies-Etude Comparative*, Ph. D. Thesis, ENPC.
192. Tsujimura T., Takao K., Sato K. (1993), 'Recent Trend of Brake Disc Material', *Japan. Rail. Eng.*, Vol. 32, No 3, Tokyo.
193. La Vie du Rail (1993), *L'Art de Faire Pencher les Trains*, November 1993.
194. Boutonnet, J.-Cl. (1993), 'Où en sont les Sybic', *RGCF*, November 1993.
195. Yamanaka, T. (1995), 'Vehicle Design Concept Towards 21th Century', *Japan. Rail. Eng.*, Vol. 32, No 4, Tokyo.

CHAPTER 13

195a. Bethge, W. (1990), 'Sicherung der Anlagen der Deutschen Bundesbahn gegen Electromagnetische Beeinflusung, *Glassers Annalen*, No 5.
196. Barwell, A.P. (1983), 'Route Survey for Electrification', *Track Course*, RIA, London.
197. Suddards, A.D. (1983), 'Electrification, Construction and Installation', *Track Course*, RIA, London.
198. Metzler, J.-M. (1990), *La Traction Electrique*, ENPC.
199. Irsigler, M. (1990), 'Guidelines for Construction and Maintenance of Overhead Line Systems on High-Capacity Lines', *Rail International*, June-July 1990.
200. Luppi, J., Lamon, J.-P. (1992), 'La Caténaire 25 KV', *RGCF*, March 1992.
201. Obermayer, H.J. (1986), *'Taschenbuch Deutsche Electrolokomotiven'*, Franck'sche Verlangschandlung, Stuttgart.
202. Jutard, M., Fitaire, M., Le Duc, E. (1989), 'Moyens d'Etude des Arcs de Rupture du Contact Pantographe - Caténaire', *RGCF*, November 1989.
203. Köck, F. (1990), 'Fahrzengdiagnose der ICE - Triebkopfe und anderer Hochgeschwindigkeitsfahrzeuge', *ETR*, No 6.
204. Kobayaski, T., Ikeda, K. (1994), 'Development of New Types of Contact Wire for High Speed Train on Shinkansen', *Japan. Rail. Eng.*, Vol. 34, No 1, Tokyo.
205. Nagosawa H., Murashima O. (1993), 'Development of High-Speed Section Insulator for Contact Line', *Japan. Rail. Eng.*, Vol. 32, No 4, Tokyo.

14 Index

acceleration, of a train 230
    resistance 225
accidents 5
active tilting technology 244
additional dynamic loading 84
adhesion, coefficient 228
    forces 227
    in relation to speed 228
aerodynamic, resistances 214
    problems in tunnels 220-222
aerotrain 23
air braking 231
alarm values of track defects 203, 204
alternative current traction (15,000 V 16 2/3 Hz) 252
alternative current traction (25,000 V 50 Hz) 252
anchors, for transverse resistance 164
    anti-creep 138
arsenic chemicals 212
asynchronous motors 267
atmoshperic pollution 256
axle 237, 238
    load 36, 37
automatic operation of switches 196

ballast, coefficient 69
    compacting 160, 161
    dimensioning 148, 149
    fatigue 144
    functions 33, 142
    geometrical characteristics 143

ballast, mechanical behaviour 144
    superelevation 159
    thickness 148-150
ballasted track 35
baseplate pads 139
benefits from a new project 183
bending moment (of track) 72, 73
bogie 83, 238, 239
    degrees of freedom 242
Boussinesq's analysis 70
braking, distance 232, 234
    electrodynamic 231
    electromagnetic 231
    electropneumatic 231
    percentage 233
    systems 231
broad gauge 36
    running resistances 219
buffers 240

cant 169, 170
    deficiency 171
    excess 171
    normal or actual value 171, 173
carbon 92
C.B.R. 56
Channel Tunnel 21, 22, 222
    geotechnical characteristics 52
characteristics of rail transport 4, 5
check rail 189
chemical composition of rails 92
choice of rail section 90

Index

choice of sleeper 116
chromium 92
circular arc 175
classification of railway subgrade 54, 55
clothoid 169
cohesion 79
collector shoes 251
combined transport 25, 26
comfort 48, 171, 230
compacting of ballast 160, 161
comparative running resistance in railways and cars 224
Computer Aided Design (Cad), applications in layout 182
concrete sleepers 122-128
conductor rail 251, 261
conical tread 39, 42, 43
consecutive curves (in layout) 176, 177, 178
constitutive law (of materials) 75, 76
containers 26
continuous-welded rail 109-115
   expansion zone 111
   force distribution 111
   length changes 112
cost-benefit method 183
cost of construction of a new railway line 49
Coulomb's law 39
couplings 240
creep 39
   of rail 138
criteria for electrification 258
critical speed of a vehicle 243
cross-sections of tracks 150-152
   with twin-block sleepers 150-152
   with monoblock sleepers 153
   with steel sleepers 151
   with timber sleepers 151
crossings 187-196
crossover 191
cubic parabola 169, 174

damping matrix 83
Dang Van's fatigue criterion 99
deceleration of a train 231
deficit of the railway undertaking 14
deformability of a sleeper 82, 122, 126, 129
degree of ballast filling 158
derailment, forces 165
   forms 165, 166
   in switches 193
   safety factor 166
design of the track-subgrade system 86
destressing of a continuous-welded rail 113
diamont crossing 190
diesel engine 247, 248
dimensioning of ballast 148, 149
Dirac's function 71
direct current traction (1500 V) 251
disk brakes 231
distribution, of train load 34
   of wheel load 81
Dormon's rule 57
double-headed rail 89
drainage devices 53, 67
Drucker-Prager plasticity criterion 77
dynamic, analysis 47, 82, 83
   impact factor 85
   stability of a vehicle 241

Eisenmann's theory 94
elastic, line (of sleepers) 82
   fastenings 133-136
elasticity modulus 79
elastoplastic analysis of the track system 80
elastoplasticity 76
electric traction, emergence of 247
   feasibility analysis 257
   principles of 250
eletrical transmission 249
eletrodynamic braking 231
eletromagnetic braking 231
eletropneumatic braking 231

embankment 62, 63
emergency values of track defects 203, 204
environment 5
European Conference of Ministers of Transport 8, 30
Eurostar 22
evolution of defects 204, 205, 206
expansion devices 114, 115

fastenings 132-137
  list of types 136
  stress and strain 137
fatigue, curve 98
  stresses in rails 100
feasibility studies 183, 257, 258
feeder conductors 258
filter 66, 142
final design 185
finite elements 74, 75
  for the study of rail 97
  in disc brake design 232
  models, results 75, 80
fishplates 107
flash-butt welding 112
forecasts for rail traffic 13
formation layer, definition and characteristics 56
  thickness 57
fouling distance 189
foundation of poles 263
Fourrier transform 72
fracture of glass windows 223
freight traffic 8, 9, 10, 27, 216, 218
frequencies of rail vibrations 84
friction angle 79
frost 54, 61, 62
  foundation thickness 61
  index 60
  protection methods 61, 62
funicular 180

gauge, definition and forms 35
gauge, permissible values 201

gas turbine locomotives 247
Gate turn off (GTO) 265
geotechnical classifications of soils 51
geotextiles 65, 66, 67
G.N.P. 8
gradients 180
granulometric composition of ballast 143
gravity resistance 225
grooved rail 89
grooves (in timber sleepers) 164

hand tools 208
hard steel rails 93
hardness of ballast 147
heavy maintenance machinery 207, 208
herbicides 212
Hertz's ellipse 94, 227
high speeds 16-20
high speed tracks (technical characteristics) 21
historical evolution of railways 1
horizontal, cracking of rail 102
  defect 200
hybrid models 77
hydrodynamic transmission 249
hydrogeological conditions 53
hydrostatic transmission 249

inertial resistance 225
initial stress method 78
interaction between rail components 209
internal discontinuities in rails 100, 101
Internal Rate of Return (IRR) 183
International Railway Institutions 28, 30
International Railways Union (UIC) 28, 30
interval between maintenance sessions 205

Kalker's theory 40
Kelvin-Voigt model 83
kinematics of the wheel 42
Klingel's analysis 41
Kronecker's delta 76

Index

land occupation 5
laying the track 153-155
layout 168, 186
Le Shuttle 22
liberalization 4, 14
lifetime of, rail 106, 107
    sleepers 119, 121, 125, 129
light maintenance machinery 208
limit conditions 75
liquidity limit 66
load gauge 43
    static 43
    dynamic 43
local distortion 201
local distortion 201
logistics 27
long-pitch corrugations 102
longitudinal, defect 200
    forces 47
    transition 180, 181
Los Angeles test 55, 146

Maison's formula 232
maintenance, coefficient 58
    of track 197-212
manganese 92
manual operation of switches 194
market requirements 8, 15
mass transit systems 25
maximum and minimum speeds in layouts 178, 179
mean settlement 198
mechanical, characteristics of materials 79
    characteristics of the rail 92
    behaviour of ballast 144
    behaviour of sleepers 122, 126, 129
    behaviour of fastenings 137
    equipment for maintenance works 207
    resistances 214
metro 25, 172, 230, 231
mesh (of the track system) 74, 75
metal wheel 40

metric gauge 36
Minden's formula 233
Miner's rule 98
mobility 5, 6, 7
monoblock, prestressed-concrete sleeper 123, 126, 129

Nadal's formula 165
narrow gauge, running resistances 219
Net Present Value Method 183
networks of railways 11
noise levels 87, 88
    damping 87
    in high speeds 88
numerical models 77

outline design 184
overturning of vehicle 166

pads 138-140
pantograph 251, 262
partial perimeter 215
passenger traffic 8, 9
passive tilting technology 244
Pedeluck's formula 233
pendolino 244
photoelasticity 97
pickaxe 208
planning, of maintenance operations 210
    of train operation 235
plastic stresses in rail 96
plasticity, criteria 76, 77
    index 55, 66
poles 259, 262
Poisson's ratio 79
positioning machine of rails 154
power of a train 215, 229
    in relation to the type of a train 230
preliminary design 184
Present Value Method 183
pressure waves in tunnels 221
private car ownership index 5, 7
privatization 4
productivity 10

profiles of rail 89, 90, 91
propelling force 40
public transport 7, 25

radius of curvature, in layout 179
   longitudinal 181
rail, fatigue 98
   defects 101, 102, 103, 104
   positioning machine 154
   section 90
   slope on sleeper 43
Rayleigh waves 86
rebounding of the wheel on the rail 165
reconfiguration of tracks (before eletrification) 261
recording methods of track defects 202
reinforced soil 64
remote control 266
resistance, aerodynamic 214
   in tunnels 221, 222
   inertial 225
   mechanical 225
   due to track curves 224
restructuring of the railways 15
rigid fastenings 133
rim (of the wheel) 237
Ro-Ro 26
rubber pads 34, 138, 139
rubber wheel 40
running resistances 214-219
   formulas of American Railways 217
   formulas of French Railways 216
   formulas of German Railways 218
   in tunnels 221

Sauthoff's formula 218
scale of layout 184, 185
scheduling, maintenance operations 208
   of a train 235
screw-type elastic fastenings 134
series excited motors 251
settlements (at rail, sleeper and subgrade levels) 80
share in the transport market 8, 9

shear, force 72, 73
   stresses 96, 100
shelling 104
shift in layout 175
shifting of the track 165
Shinkansen 2, 16, 17
shoe brakes 231
short-pitch corrugations 102
shuttle 22
slab track 35, 130, 132
slab track, geometrical characteristics 132
sleeper, concrete 122-128
   deformability 82, 122, 126, 129
   functions 116
   spacing 38
   steel 117, 118
   timber 119, 122
slip 191
specific, power 230, 254
   resistance 216
   traction force 226
spectral analysis 202
speed in relation to curvature 179
spray trains 212
springs 239, 240
spring-type elastic fastenings 134
sprung masses 47
staking the layout 185
standard gauge 36
starting force 226
steam 2, 245, 246
   locomotive 246
steel sleepers 117, 118
strain models 77
stray currents 131
strength of ballast 147
stress-strain, relationship 75, 76
stress, in rail 93, 94, 95, 96
   in the subgrade 80
   models 77
   under the sleeper 130
Strahl's formula 218
stiffness matrix 78

Index

stochastic phenomena 166
structure gauge 43
subballast 34, 142
subgrade, elasticity 70
    fatigue 59
    mechanical characteristics 55, 56
    plastic deformations 59, 60
    stresses 80, 57
substations 250, 264, 265
suburban trains, running resistance 217
superelevation ramp 177
suspension, of transmission line 260
    of vehicles 239
switches 187-196
    for high speeds 193, 194
switch rail, cross-section 189
synchronous motors 266

tache ovale 102
Talgo 244
T.G.V. 3, 17, 21, 214, 215
technical considerations for maintenance 211
thermit welding 113
thickness of ballast 148-150
thyristors 265
ties (see sleepers)
tilting vehicles 171, 243, 244
timber sleepers 119-122
tongue rail 189
track, coefficients 68, 69
    construction program 155
    defects 197, 199, 200
    gauge 35, 36
    index 68
    laying 153-155
    reaction coefficient 68
    stiffness 68
    system 23
traction systems 251
traffic load 37
tramway 25, 89
transition curve 172-174
transmission, line 259
    systems (for diesel operation) 249

transverse, acceleration 171
    defect 200
    forces 46
    gradient 53, 67
    resistance 157
    track forces 156
trench 62, 63
tunnel cross-section 223
turnout, components 189
    geometrical characteristics 188
    forms 190
    running speed 192
    layout 195
twin-block reinforced-concrete sleeper 123-126
tyre 237

UIC (International Railways Union) 28
UIC classification of lines (in relation to traffic load) 38
UIC rail profiles 91
underfilling 207
unsprung masses 47

variable stiffness method 78
vertical transition 180, 181
vibrations 86, 87, 88
Vignole-type rail 89, 90
viscoelastic behaviour 83
Von Mises plasticity criterion 77

wear of rail 105, 106
weed control 212
welding of rail 112, 113
wheel 38, 236
    base 237
    diameter 228, 236
width of rolling stock 43
Wöhler's curve 98
wood quality of timber sleepers 119, 163

X 2000 train 244

Zimmermann's method 71